Analog Circuits and Signal Processing

The *Analog Circuits and Signal Processing* book series, formerly known as the *Kluwer International Series in Engineering and Computer Science*, is a high level academic and professional series publishing research on the design and applications of analog integrated circuits and signal processing circuits and systems. Typically per year we publish between 5-15 research monographs, professional books, handbooks, and edited volumes with worldwide distribution to engineers, researchers, educators, and libraries. The book series promotes and expedites the dissemination of new research results and tutorial views in the analog field. There is an exciting and large volume of research activity in the field worldwide. Researchers are striving to bridge the gap between classical analog work and recent advances in very large scale integration (VLSI) technologies with improved analog capabilities. Analog VLSI has been recognized as a major technology for future information processing. Analog work is showing signs of dramatic changes with emphasis on interdisciplinary research efforts combining device/circuit/technology issues. Consequently, new design concepts, strategies and design tools are being unveiled. Topics of interest include: Analog Interface Circuits and Systems; Data converters; Active-RC, switched-capacitor and continuous-time integrated filters; Mixed analog/digital VLSI; Simulation and modeling, mixed-mode simulation; Analog nonlinear and computational circuits and signal processing; Analog Artificial Neural Networks/Artificial Intelligence; Current-mode Signal Processing; Computer-Aided Design (CAD) tools; Analog Design in emerging technologies (Scalable CMOS, BiCMOS, GaAs, heterojunction and floating gate technologies, etc.); Analog Design for Test; Integrated sensors and actuators; Analog Design Automation/Knowledge-based Systems; Analog VLSI cell libraries; Analog product development; RF Front ends, Wireless communications and Microwave Circuits; Analog behavioral modeling, Analog HDL.

Shiva Maleki Varnosfaderani • Nabil J. Sarhan •
Mohammad Alhawari

Hardware Software Co-design of Epileptic Seizure Prediction Systems

Springer

Shiva Maleki Varnosfaderani
Wayne State University
Detroit, MI, USA

Nabil J. Sarhan
Wayne State University
Detroit, MI, USA

Mohammad Alhawari
Wayne State University
Detroit, MI, USA

ISSN 1872-082X ISSN 2197-1854 (electronic)
Analog Circuits and Signal Processing
ISBN 978-3-031-95356-9 ISBN 978-3-031-95357-6 (eBook)
https://doi.org/10.1007/978-3-031-95357-6

This Springer imprint is published by the registered company Springer Nature Switzerland AG
The registered company address is: Gewerbestrasse 11, 6330 Cham, Switzerland

To my dear family—my parents, Ahmad and Rezvan; my sisters, Mina, Sima, and Shirin; and my brother, Farzad—for your unconditional love, strength, and belief in me, even across the miles. To Dr. Alhawari, for your generous mentorship, guidance, and support throughout this journey. To all patients living with epilepsy—may this work be a small step toward a future filled with healing, dignity, and hope. And to all kind-hearted people who make life more beautiful through their kindness—this book is also for you.

Preface

Epilepsy, a complex and often unpredictable neurological disorder, affects millions of people globally, profoundly impacting their daily lives and medical treatment. Despite advancements in medicine, accurately predicting epileptic seizures remains a significant challenge. With the rapid evolution of deep learning and machine learning, new technologies now offer the potential to revolutionize epilepsy prediction, enhancing patient care and improving quality of life. This book is a comprehensive effort to bridge the gap between medical expertise and engineering innovations, bringing together insights from both fields to tackle the problem of epileptic seizure prediction.

Why this book is essential: This book provides a unique combination of medical and technical perspectives, making it an invaluable resource for doctors, engineers, researchers, and data scientists. Whether you're a medical professional seeking to understand how machine learning can be applied to patient care or an engineer interested in real-world medical challenges, this book is designed to guide you through the complexities of epilepsy prediction using deep learning techniques.

By reading this book, you will:

- Gain a comprehensive understanding of epilepsy, including its types, symptoms, and the challenges faced by patients and clinicians.
- Learn about machine learning models used in epilepsy prediction, with detailed comparisons of their performance and key components.
- Understand how to analyze and clean EEG data, enabling you to build more accurate seizure prediction systems.
- Explore advanced denoising techniques and preprocessing methods for EEG data to improve model performance.
- Discover efficient hardware solutions for low-power, real-time seizure prediction systems, crucial for wearable or implantable medical devices.
- Benefit from practical insights based on extensive analysis of real-world EEG datasets, including Melbourne, CHB-MIT, and European Epilepsy datasets, providing a thorough comparison of data sources and their relevance to seizure prediction. What sets this book apart is its in-depth analysis of how various

factors, such as seizure onset zone, patient characteristics, and seizure types, influence the performance of prediction models. These patient-specific insights ensure that the solutions provided are not just theoretical but can be applied to real clinical settings. We examine how the proposed models perform under varying conditions, considering critical factors like electrode placement, signal quality, and individual patient traits, offering readers a holistic view of the challenges and opportunities in seizure prediction.

The book is structured to provide both foundational knowledge and advanced technical solutions:

Chapter 1 says the groundwork with a detailed exploration of epilepsy, its challenges, and the tools used for diagnosis and monitoring.
Chapter 2 reviews the history of epilepsy prediction models, offering a comprehensive comparison of past and current machine learning techniques.
Chapter 3 provides an in-depth analysis of the most popular public and private EEG datasets used for seizure prediction, helping readers understand how to select and work with these datasets.
Chapter 4 focuses on advanced signal processing techniques, evaluating various denoising methods to enhance the quality of EEG data.
Chapter 5 introduces a computationally efficient LSTM-based model for automatic patient-specific seizure prediction, evaluating how factors like seizure type and dataset characteristics affect prediction accuracy.
Chapter 6 proposes an innovative power management model for hardware-based epilepsy prediction systems, designed to optimize battery life without compromising performance.

Why should you buy this book? This book offers a rare combination of cutting-edge machine learning techniques and practical medical insights, making it essential for anyone interested in the future of epilepsy treatment. Whether you're a medical professional aiming to incorporate advanced technologies into patient care or an engineer developing new solutions for healthcare, this book provides the tools and knowledge you need to create impactful systems that can predict seizures more accurately and efficiently.

By the end of this book, readers will be equipped with a deeper understanding of how machine learning can be applied to epilepsy prediction, how data should be handled for better outcomes, and how patient-specific characteristics can be incorporated into predictive models to enhance performance. This knowledge will not only advance research in the field but also lead to the development of more effective, patient-centered solutions.

Take the next step toward improving the lives of epilepsy patients with insights that blend medical science and engineering innovation. This book is your guide to making a tangible difference.

Detroit, MI, USA

Shiva Maleki Varnosfaderani
Nabil J. Sarhan
Mohammad Alhawari

Acknowledgments I would like to express my deepest gratitude to my academic advisor, Dr. Alhawari, for his invaluable mentorship, encouragement, and unwavering support throughout this journey. His insights and guidance have played a pivotal role in shaping this work. I am truly grateful for the opportunity to join his research group, where I have grown both professionally and personally.

I am sincerely thankful to Dr. Sarhan, who contributed significantly to the research and development of this book as my co-advisor.

A heartfelt thanks to my dear family-my father, Ahmad; my mother, Rezvan; my sisters, Mina, Sima, and Shirin; and my brother, Farzad. Thank you for always standing by my side, for your endless love, support, and encouragement. Your presence has given me strength through every step of this journey. Your support and courage inspire me every day. I am forever grateful for your love and for always believing in me. Even though we are miles apart, distance means nothing when our hearts are always with each other.

I would also like to thank my friends for their kindness, encouragement, and support along the way.

Finally, I am grateful to the team at Springer Nature for their assistance in preparing this book for publication and for their dedication to advancing academic and scientific knowledge.

I hope this book serves as a valuable reference for researchers and medical professionals who care deeply about patients with epilepsy, and that it contributes, in some way, to making their lives healthier and happier. I believe that with kindness in our hearts, we can make the world a better place for everyone.

Above all, I thank God for placing me on this path, for being my source of strength and hope, and for surrounding me with people who lifted me up along the way.

Competing Interests The authors have no competing interests to declare that are relevant to the content of this manuscript.

Contents

Acronyms

Adam	Adaptive Moment Estimation
AED	Anti-epileptic Drug
AES	American Epilepsy Society
AI	Artificial Intelligence
AUC	Area Under the Curve
BRAM	Block RAM
BSNN	Bit-Serial Neural Network
BSS	Blind Source Separation
CAR	Common Average Reference
CGRNN	Convolutional Gated Recurrent Neural Network
CP	Complex Partial
CT	Computed Tomography
CNN	Convolutional Neural Networks
DCGAN	Generative Adversarial Networks
DFT	Directed Transfer Function
DL	Deep Learning
DNN	Deep Neural Network
DPU	Data Processing Unit
EDA	Electrodermal Activity
EEG	Electroencephalogram
EMD	Empirical Mode Decomposition
FF	Flip Flop
FN	False Negatives
FP	False Positives
FSM	Finite State Machine
GAN	Generative Adversarial Network
GCN	Graph Convolutional Network
GTCS	Generalized Tonic-Clonic Seizure
iEEG	Intracranial EEG
ILAE	International League Against Epilepsy
IoT	Internet of Things

IWSP	International Workshop on Seizure Prediction
LOOCV	Leave-One-Out Cross-Validation
LSTM	Long Short-Term Memory
MAC	Multiply-Accumulate
MD	Minimum Distance
ML	Machine Learning
MRI	Magnetic Resonance Imaging
NMSE	Normalized Mean Squared Error
PET	Positron Emission Tomography
RNS	Responsive Neurostimulation System
ROC	Receiver Operator Characteristic
sEEG	Scalp EEG
SG	Secondarily Generalized
SNR	Signal-to-Noise Ratio
SOZ	Seizure Onset Zone
SP	Simple Partial
SPECT	Single Photon Emission Computed Tomography
STFT	Short-Time Fourier Transform
SUDEP	Sudden Unexpected Death in Epilepsy
SURE	Stein's Unbiased Risk Estimate
SVM	Support Vector Machine
T	Threshold
TN	True Negatives
TP	True Positives
U	Unclassified
UT	Electrodes-of-Interest (EOI)
ViT	Vision Transformer
VNS	Vagus Nerve stimulation
WHO	World Health Organization
WT	Wavelet Transform

Chapter 1
Epilepsy Fundamentals: Challenges and Perspectives

1.1 What Is Epilepsy?

The brain is made up of many billions of neurons. Neurons create networks with other neurons and communicate with each other through chemical and electrical impulses. When there is a sudden abnormal electrical activity that briefly disrupts normal brain function, a seizure happens. The majority of seizures are caused by a very small number of abnormal neurons that affect nearby or networked neurons. As more and more neurons in the network become involved during a seizure, the pattern of hypersynchrony is produced. This abnormal propagation happens due to insufficient inhibition, excessive excitation, or both within the neural network. This atypical neuronal hypersynchrony may develop or be congenital at any point in life [1, 2]. The tendency to experience repeated, unprovoked seizures is known as epilepsy [2, 3]. Epilepsy is described as a brain disorder that is characterized by a persistent propensity to have epileptic seizures as well as the associated neurological, cognitive, psychological, and social effects. This definition highlights the fact that epilepsy consists not only of the neurobiology of seizures but also of the associated neuropsychosocial comorbidities. After a single seizure, epilepsy can be diagnosed if the patient has a greater than 60% probability of experiencing another unprovoked seizure and the seizure was unprovoked (i.e., unrelated to drugs, alcohol, hyponatremia, or glucose abnormalities) [2].

The International League Against Epilepsy (ILAE) defines epilepsy as any of the following: (1) at least two unprovoked (or reflex) seizures that happen more than 24 hours apart; (2) one unprovoked (or reflex) seizure and a probability of additional seizures that occur within the next 10 years that is comparable to the general recurrence risk (at least 60%) after two unprovoked seizures; and (3) diagnosis of an epilepsy syndrome [4]. Nonetheless, the ILAE Epidemiology Commission recommends that epilepsy be classified as two or more unprovoked seizures that happen at least 24 hours apart in order to facilitate population-based

S. M. Varnosfaderani et al., *Hardware Software Co-design of Epileptic Seizure Prediction Systems*, Analog Circuits and Signal Processing,
https://doi.org/10.1007/978-3-031-95357-6_1

research [5]. Epilepsy is redefined as a "disease" and not a "disorder" by the ILAE to emphasize its importance and impact, although some people might be worried that using disease carries more stigma diseases like cancer, heart disease, and so forth [6]. The patients having experience in convulsive movements and loss of consciousness are under serious physical injury [7]. Patients who had convulsive movements and loss of consciousness are at a greater risk of sudden unexpected death in epilepsy (SUDEP), which is a leading cause of death in patients with uncontrolled seizures [8]. SUDEP is defined as sudden, unexpected, nontraumatic, nondrowning, witnessed, or unwitnessed death in epileptic patients, with or without documented status epilepticus, in which postmortem examination fails to identify a toxicologic or anatomic cause of death [9]. Most of the time, a seizure causes SUDEP; thus it seems sense to assume that seizures might cause changes in the cardiorespiratory system. Due to uncontrolled seizures, epileptic patients are not able to obtain meaningful employment and are restricted from operating a motorized vehicle [7]. Also, according to the strong evidence, patients with long-standing uncontrolled epilepsy are at the risk for anxiety, suicide, and other psychiatric disorders having negative effects on the quality of patients' lives and their families' lives. In addition, because of the healthcare cost, they also represent an economic burden to their family members [7].

Epilepsy is a neurological condition that can cause unconsciousness, jerking of some body parts, and, in severe cases, uncontrollable whole-body convulsions. People of various ages might be affected by this illness. The patient's quality of life may be significantly impacted, and further social and economic repercussions may arise. Epilepsy may also result in early death. Epilepsy is a seizure disorder that is usually diagnosed after at least two seizures that are not caused by other medical disorders [3]. Epileptic seizures may be caused by a brain injury or a genetic predisposition, although the reason is frequently unclear.

1.2 Population of Epileptic Patients

Epilepsy is a widespread neurological condition that affects individuals of all ages, races, social classes, and geographic locations. It ranks as the fourth most common neuralgic condition. Over 65 million individuals worldwide, including 3.4 million Americans, have epilepsy, according to the World Health Organization (WHO) [10]. Approximately 1 in 26 Americans will experience epilepsy at some point in their lives. There is a 3% lifetime prevalence rate. People of all ages and backgrounds are affected, but it usually begins before the age of one year and develops once again around the age of 50, peaking at 75 [11]. With the right care, though, about 70% of the patients may experience no seizures. Thirty percent of individuals may develop medication-resistant epilepsy, defined as failing to achieve prolonged seizure freedom after two sufficient trials with well-tolerated, carefully selected, and administered antiepileptic drug (AED) regimens [4].

A meta-analysis and comprehensive review of incidence studies showed a pooled incidence rate of 61.4 per 100,000 person-years for epilepsy. The incidence was greater in low- and middle income countries (139.0) compared to high income countries (48.9) [12]. This can be interpreted in conjunction with the fact that low- and middle-income countries have a higher incidence of traumatic brain injury and central nervous system infections, as well as a distinct composition of people at risk and increased exposure to perinatal risk factors. Additionally, among the same population, those with varying ethnic backgrounds and those from the lowest socioeconomic strata in high-income countries had greater incidences of epilepsy [13]. Differences may also be caused by methodological problems, such as the removal of isolated and acute symptomatic seizures in some studies and stricter case verification [9].

Depending on the number of seizures upon diagnosis, the local distribution of risk and etiologic variables, and whether cases in treatment are included in the lifetime prevalence calculation or just active epilepsy is taken into account, the prevalence of epilepsy varies greatly between countries. The total lifetime prevalence of epilepsy was 7.60 per 1,000 and was greater in low- and middle-income countries (8.75 per 1,000) than in high-income countries (5.18 per 1,000) [12]. 6.38 cases of active epilepsy per 1,000 people was the point prevalence. Active epilepsy had a median point prevalence of 6.68 in low- and middle-income countries and 5.49 in high-income countries [9]. Prevalence estimates in certain populations also differ and are often greater in members of particular ethnic groups [14], those with poor health, and participants from socially underprivileged backgrounds [15]. In addition to problems with the research design, the study population's demographic makeup, the frequency of environmental risk factors, and the standard of health care can all be connected.

Men are a little more likely than women to experience epilepsy in terms of both prevalence and incidence [12]. Some locations may have varying rates of the most frequent risk factors, and women may hide the disease for social or cultural reasons, which might account for the discrepancy [16]. Epilepsy is more common in the youngest and oldest age groups [12], with estimates of 86 cases per 100,000 people in a well-defined population in the first year of life, a trend lower to roughly 23–31 cases per 100,000 in the 30–59 age group, and a subsequent rise to 180 cases per 100,000 in the over 85 age group [12]. The incidence of epilepsy in children is highest level in the first year of life, and by the end of the 10th year of life, it declines to adult levels [17]. Children in low- and middle-income countries have epilepsy at a higher peak; this might be due to the country's demographic makeup and a lack of awareness of the condition in older people.

According to the 2016 Global Burden of Disease Collaborators, epilepsy affects over 46 million people globally, making it a significant portion of the illness burden [18]. Whereas rates of epilepsy prevalence and incidence are greater in low- and middle-income countries than those in high-income countries, about 80% of epileptics lives there [19]. Less access to healthcare, a higher frequency of injuries, and different causes are probably the reasons for the disparities. Depending on the location, epilepsy ranked second to eighth among all neurological illnesses in terms

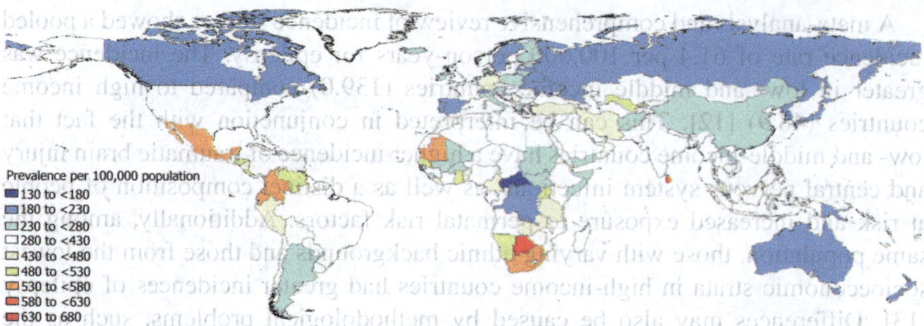

Fig. 1.1 Age-standardized prevalence (×100,000) of idiopathic epilepsy by country, 2016 [18]

of age-standardized DALY rates by Global Burden of Disease region in 2016. As can be seen in Fig. 1.1, Southeast Asia, central and Andean Latin America, eastern, western, and southern sub-Saharan Africa, and central Asia had the greatest rates of idiopathic epilepsy (i.e., caused by a genetic cause or when diagnostic evaluation failed to identify a causal component).

1.3 Brain Anatomy, Symptoms of Seizures, and Surgical Intervention

In Fig. 1.2, the brain anatomy is displayed. As can be seen, each area of the brain plays a critical role in controlling various bodily functions. During an epileptic seizure, abnormal electrical activity disrupts these functions depending on where the seizure begins, known as the seizure onset zone. Figure 1.3 depicts common symptoms and signs of seizures. Seizures can arise from any of these regions and, depending on the onset zone, lead to a wide variety of symptoms. Understanding the relationship between brain regions and their associated functions is essential for predicting and managing the symptoms of epilepsy, as well as identifying areas for potential surgical intervention, especially in drug-resistant cases.

The frontal lobe is responsible for higher cognitive functions such as reasoning, problem-solving, and emotional regulation. It also controls voluntary movements, speech, and memory. Seizures originating in the frontal lobe can manifest as motor seizures, often presenting as sudden jerking movements on one side of the body, known as Jacksonian seizures. These seizures may also cause speech disturbances, where patients are unable to speak or experience speech arrest, along with emotional or behavioral changes, such as emotional outbursts or repetitive, purposeless movements (automatisms). For drug-resistant cases, a frontal lobectomy may be performed. This surgical procedure removes the seizure focus while preserving critical functions like movement and speech, and is particularly effective for patients with focal motor seizures or complex partial seizures originating in this region.

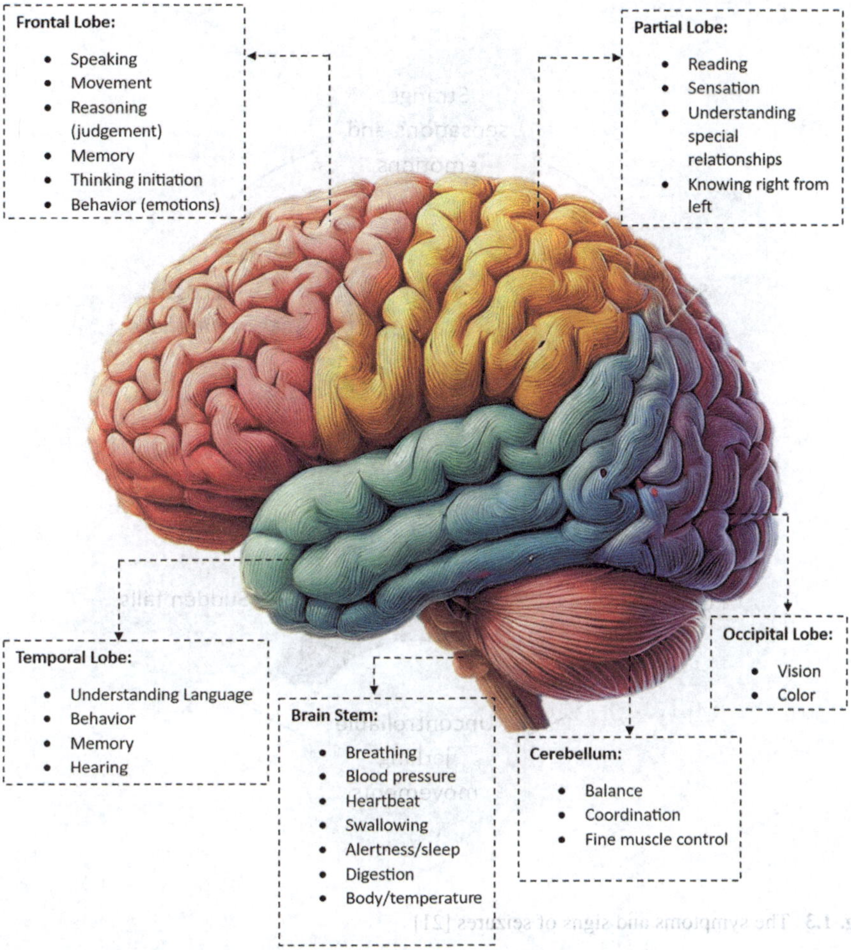

Fig. 1.2 Human brain's lobes and their associated functions [20]

The temporal lobe governs memory, language comprehension, and auditory processing, and is integral to emotions and behavior. Temporal lobe seizures often begin with an aura, which may involve strange sensations such as unusual smells, or a sense of impending doom. Memory lapses and confusion are common during and after these seizures, and patients may exhibit complex partial seizures, characterized by repetitive behaviors (e.g., lip-smacking or hand gestures), despite appearing conscious. In cases of drug-resistant epilepsy, temporal lobectomy is a common and successful surgical option. This procedure involves removing the seizure focus, often in the hippocampus or amygdala, areas critical to memory and emotions. Many patients experience a significant reduction in seizures or complete seizure freedom following surgery, especially those with medial temporal lobe epilepsy.

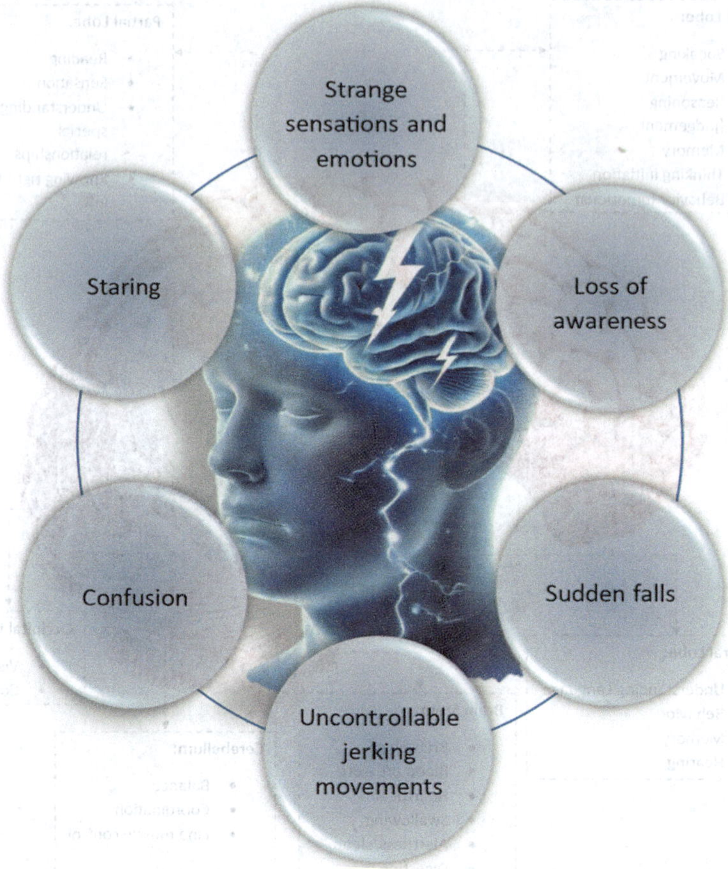

Fig. 1.3 The symptoms and signs of seizures [21]

The parietal lobe processes sensory information from the body and is involved in spatial awareness and navigation. Seizures originating here often lead to sensory disturbances like tingling, numbness, or burning sensations in different parts of the body, referred to as paresthesia. Other symptoms include spatial disorientation, where patients have difficulty understanding their surroundings, and altered body perceptions, where a limb may feel as though it is missing or has changed size. Parietal lobe resection may be performed to treat drug-resistant epilepsy in this region, which helps alleviate sensory-related seizures and associated symptoms.

The occipital lobe is the brain's center for visual processing, interpreting information from the eyes. Seizures in this area often cause visual disturbances, such as hallucinations, flashes of light, or even temporary blindness. Patients may also experience ocular symptoms, where objects appear distorted or involuntary eye movements occur. Occipital lobe resection can be considered for patients with drug-resistant epilepsy originating in this area, though the procedure is performed with caution due to the critical nature of visual processing.

The cerebellum plays a key role in maintaining balance, coordination, and fine motor control. Seizures affecting the cerebellum result in loss of coordination, balance problems, and difficulty walking or standing. Patients may also experience tremors, with involuntary shaking of the hands or legs. Surgical intervention in the cerebellum is rare due to its essential role in motor control. Treatment for cerebellar seizures usually focuses on medication and noninvasive approaches rather than surgery.

Finally, the brain stem regulates vital bodily functions such as breathing, heart rate, and blood pressure. Seizures that arise in this region can lead to autonomic symptoms, including disturbances in breathing, heart rate, and blood pressure. Patients may also experience nausea, vomiting, or excessive sweating. Due to the critical role of the brain stem in maintaining consciousness, seizures here often result in loss of consciousness or fainting. Direct surgery on the brain stem is extremely rare, given its essential functions. Instead, treatments such as vagus nerve stimulation (VNS), which involves implanting a device to modulate brain activity via the vagus nerve, may be explored for patients with seizures originating from this area.

By understanding the functions of different brain regions and how seizures manifest when they originate in these areas, clinicians can better identify appropriate treatments. Surgical intervention, when viable, can provide significant relief for patients with drug-resistant epilepsy, particularly when the seizure onset zone is well defined and surgically accessible. Every area of the brain is in charge of a certain function. Patients have different symptoms during an epileptic seizure based on the part of the brain that is affected. Therefore, comprehending the function of the brain can aid in determining the likelihood of symptoms that patients may encounter.

1.4 Methods to Diagnosis the Epilepsy

While most epileptic patients recover well with appropriate care, others go on to suffer from severe mental problems. Consequently, ongoing medical care might be required. Patients' health can be improved by an early diagnosis [22]. From the clinical point of view, we might have epilepsy if two or more impulsive seizures take place. When epilepsy is the cause of a seizure, it is very beneficial to recognize the condition at its start and beginning of antiepileptic therapy as soon as possible to improve the quality of life and safety of epileptic individuals. Physical dangers, including drowning (if the seizure occurs while swimming or taking a bath), burns, and head injuries, are mostly caused by the unpredictable nature of the seizures [23]. Epileptic seizures are diagnosed using a variety of techniques. These techniques include genetic testing, positron emission tomography (PET), magnetic resonance imaging (MRI), computed tomography (CT) scans, electroencephalogram (EEG), and single-photon emission computed tomography (SPECT). Basic blood tests are also carried out since they can be useful in determining the cause of metabolic and toxic encephalopathies [24]. According to studies, the two main methods used to

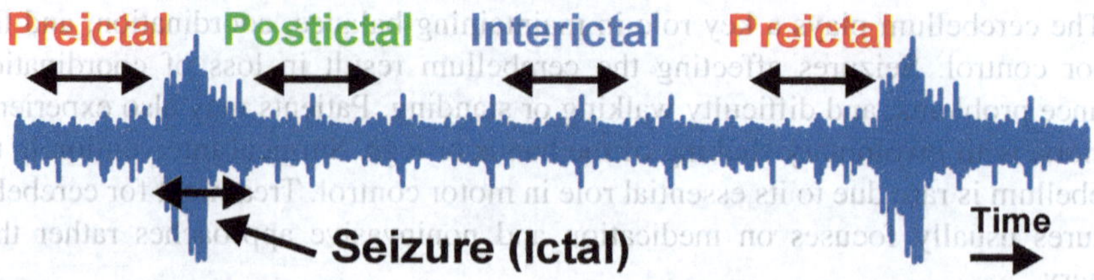

Fig. 1.4 Various stages of EEG signal in epileptic seizures [26]

diagnose epileptic seizures are MRI and EEG. Additional methods can even reveal misleading negative findings while aiding in diagnostic confirmation [22].

In order to make a clinical diagnosis of epilepsy, a thorough medical history, neurological exams, blood tests, and occasionally testing on cerebrospinal fluid are required to check out any possible linked causes, such as previously described biochemical abnormalities.

The electroencephalogram (EEG) is a quick and fairly cheap way to examine and research brain electrical activity. The procedure involves inserting electrodes either on the scalp (known as scalp EEG (sEEG) recording) or inside the skull (known as intracranial EEG (iEEG) recording). Epileptic seizures can be detected using EEG signals by identifying certain abnormal brain activities associated with epileptic seizures, such as sharp spikes. As shown in Fig. 1.4, the EEG signal can be classified into four main categories: *preictal*, *seizure or ictal*, *postictal*, and *interictal*. Preictal is the time before seizure occurrence which lasts around 15 minutes to one hour, whereas postictal is the time right after a seizure, and interictal is the time between preictal and postictal stages. Ictal refers to the phase when a seizure takes place [25].

While CT scans are useful as well, only 30% of patients have their focal lesions detected. When choosing a neurosurgical procedure for therapy, neuroimaging is crucial to the assessment of epilepsy. Improvements in MRI technology, acquisition techniques, and methodologies have led to significant advancements in brain imaging. The primary neuroimaging method used to identify an epileptogenic lesion is structural MRI [27]. However, 15% to 30% of individuals with refractory focal epilepsy may get erroneous negative results.

The most recent MRI hardware and procedures are employed if a patient has no obvious lesions on their MRI, which may reveal some minute anomaly. But because there is a chance of a false-positive outcome, this needs to be done very carefully [28]. There are a few justifications for using MRI to look into the medical condition. Prior to surgery, MRI can assist in evaluating the problem and identifying the reason [29].

Functional imaging investigations like PET and SPECT help to determine the epileptogenic zone and provide preoperative assessment. Genetic testing is utilized to determine the cause of some epilepsies. There are other restrictions, nevertheless, such as the comparatively expensive price and limited availability [30]. Techniques such as PET imaging and SPECT imaging are used to identify the region of the brain that initiates seizures, particularly in patients with focal epilepsy and a normal

MRI, in patients with numerous abnormalities, or in patients whose MRI and EEG show inconsistent results. SPECT imaging also provides information on variations in cerebral perfusion [31].

1.5 Classification of Epilepsy and Seizures

The categorization of seizure types is crucial in identifying the cause, the most effective procedure for therapy, and the overall prognosis. The appearance of the seizure and the areas of the cerebral cortex involved in the seizure are used to categorize seizures. To assist doctors in diagnosing and treating individuals with epilepsy, ILAE defined categories of epilepsy based on the seizure onset zone in 1981. In 2017, they updated this classification (Fig. 1.5). Focal epilepsy, which is limited to one hemisphere of the brain, generalized epilepsy with apparent clinical or EEG onset in both hemispheres of the brain, and unknown epilepsy are the three categories of epilepsy in the new ILAE categorization system [32, 33]. More details are obtained in the following.

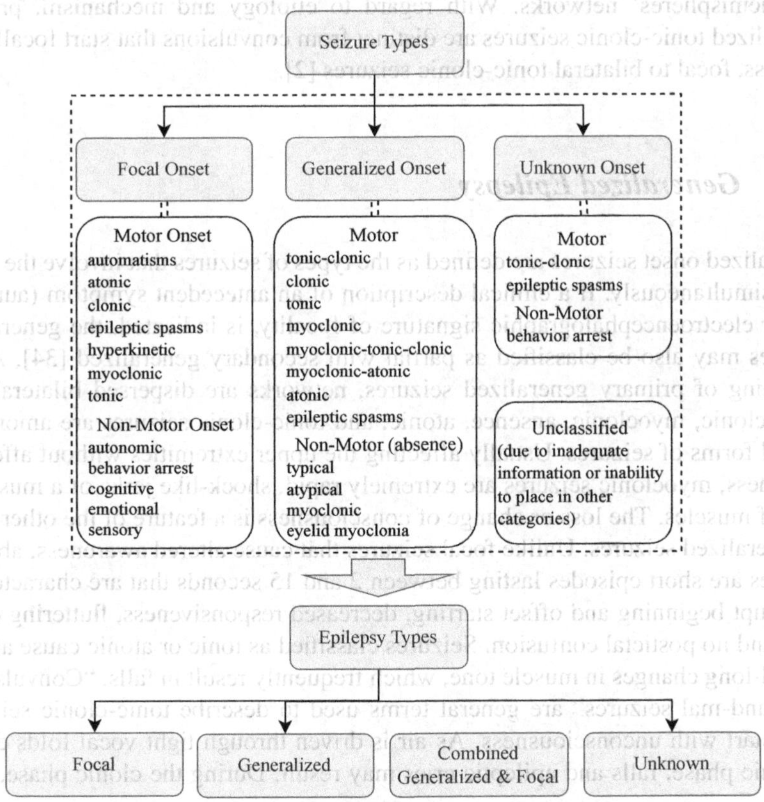

Fig. 1.5 The expanded version of ILAE 2017 classification of seizure types [2]

1.5.1 Focal Epilepsy

Partial or focal epilepsy is defined as seizures that start in a specific region of the brain and are limited to one hemisphere of the brain. In the older classification by ILAE, this type of seizure is further classified as simple partial seizures (consciousness remains unchanged) and complex partial seizures (consciousness changes) [34]. There may be complete awareness and consciousness preservation during a focal seizure. The patient is conscious, responsive, and able to remember the seizure. Aura or simple partial is similar to this kind of seizure. Changes in responsiveness and memory can either occur during focal seizures, which can start with altered consciousness or develop to it. In the older categorization system, this kind of seizure was known as a complex partial seizure. It is now known as a focal seizure with impaired consciousness. Automatisms involving the eyes (e.g., blinking), mouth (e.g., lip-smacking and chewing), hands (e.g., fumbling and picking), vocalizations (e.g., grunts and repeating words or phrases), or more complex acts (e.g., walking and attempting to use a cellphone) are common during focal seizures with impaired awareness. After a short period of disorientation and exhaustion, this kind of seizure typically lasts between 30 seconds and two minutes. Tonic-clonic seizures can result from a focal seizure that spreads across both hemispheres' networks. With regard to etiology and mechanism, primary generalized tonic-clonic seizures are distinct from convulsions that start focally and progress, focal to bilateral tonic-clonic seizures [2].

1.5.2 Generalized Epilepsy

Generalized onset seizures are defined as the types of seizures that involve the entire brain simultaneously. If a clinical description of an antecedent symptom (aura), or a clear electroencephalographic signature of focality, is indicated, the generalized seizures may also be classified as partial with secondary generalized [34]. At the beginning of primary generalized seizures, networks are dispersed bilaterally. A tonic, clonic, myoclonic, ansence, atonic, and tonic-clonic seizures are among the several forms of seizures. Usually affecting the upper extremities without affecting awareness, myoclonic seizures are extremely rapid, shock-like jerks of a muscle or a set of muscles. The loss or change of consciousness is a feature of the other types of generalized seizures. Unlike focal seizures that cause altered awareness, absence seizures are short episodes lasting between 2 and 15 seconds that are characterized by abrupt beginning and offset starting, decreased responsiveness, fluttering of the eyes, and no postictal confusion. Seizures classified as tonic or atonic cause abrupt, second-long changes in muscle tone, which frequently result in falls. "Convulsions" or "grand-mal seizures" are general terms used to describe tonic-clonic seizures. They start with unconsciousness. As air is driven through tight vocal folds during the tonic phase, falls and epileptic cries may result. During the clonic phase, there is jerking of the lower and upper limbs, biting of the tongue, foaming of the mouth, and incontinence. A tonic-clonic seizure leaves the patient disoriented and listless; when awareness returns, the patient may become agitated [2].

1.6 Useful Software for Epileptic Seizure Applications

A non-exhaustive collection of commercial and academic software programs that include tools for localizing EEG sources are obtained in Table 1.1 [35]. It is essential that the user understands how the various processing stages are implemented,

Table 1.1 A partial list of commercial and academic software programs that provide EEG source localization techniques

Name	Website	Inverse models
Academic software packages		
Brainstorm	https://neuroimage.usc.edu/brainstorm	Dipole modeling, Beamformer, sLORETA, dSPM
Cartool	https://sites.google.com/site/cartoolcommunity/	Minimum Norm, LORETA, LAURA, Epifocus
EEGLab	https://sccn.ucsd.edu/eeglab/index.php	Dipole modeling
Fieldtrip	http://www.fieldtriptoolbox.org/	Dipole modeling, Beamformer, Minimum Norm
LORETA	http://www.uzh.ch/keyinst/loreta.htm	LORETA, sLORETA, eLORETA
MNE	https://martinos.org/mne/stable/index.html	MNE, dSPM, sLORETA, eLORETA
NUTMEG	https://www.nitrc.org/projects/nutmeg	Beamformer
SPM	https://www.fil.ion.ucl.ac.uk/spm/	dSMP
Commercial software packages		
BESA	https://www.besa.de/products/besa-research/besa-research-overview/	Dipole modeling, RAP-MUSIC, LORETA, sLORETA, LAURA, SSLOFO
BrainVision Analyzer	https://www.brainproducts.com/	LORETA
BrainVoyager	https://www.brainvoyager.com/	Beamformer, Minimum Norm, LORETA, LAURA
GeoSource	https://www.usa.philips.com/healthcare/solutions/neuro/neuro-research-applications	
CURRY	https://compumedicsneuroscan.com/source-reconstruction/	Dipole modeling, MUSIC, Beamformer, Minimum norm, sLORETA, eLORETA, SWARM

regardless of the program being utilized. With the current attempts to put up best practice recommendations for reporting EEG/MEG investigations, it is crucial to guarantee reproducibility and replicability by documenting the information about the procedures taken in the various software packages and providing access to this information in publications [35].

Chapter 2
Machine Learning for Epileptic Seizure Prediction

2.1 Research-Based Epileptic Seizures

Epileptic-based research can be divided into four main categories: (1) antiepileptic medications, (2) epilepsy localization and surgery, (3) epilepsy detection, and (4) epilepsy prediction (Fig. 2.1). Antiepileptic medications are currently the basis of epilepsy therapy. Unfortunately, despite receiving therapy, over 30% of people with epilepsy continue to experience seizures [36]. The other 70% of patients who react to antiepileptic drugs experience unfavorable side effects, such as nausea, fatigue, dizziness, or impaired vision. When drugs are unable to manage seizures, epilepsy surgery may be a possibility. Every patient with drug-resistant epilepsy should be assessed for potential surgical treatment. Level 1 evidence points to the superiority of epilepsy surgery over medication treatment for individuals who qualify for temporal lobectomy. Just 8% of individuals receiving medicinal treatment had no seizures after a year, compared to 58% of surgically treated patients [37]. Though MRI-guided laser ablation is becoming more popular, open excision of the epileptogenic cerebral cortex remains the most common surgical method. It has the benefit of just requiring a burr hole, as opposed to a craniotomy. Additional surgical techniques include corpus callosotomy, vagal nerve stimulation, thalamic stimulation, and responsive neurostimulation (such as NeuroPace) [2]. This surgical technique involves resecting a portion of the brain that causes seizures, which helps to halt seizures or lessen their frequency. However, it might come with significant consequences, such as paralysis, stroke, vision damage, and memory and language issues [29, 38]. Four categories may be created from the operation results: Engle Class I (no seizures after surgery), Engle Class II (rare debilitating seizures initially, but rare seizures now), Engle Class III (worthwhile improvement), and Engle Class IV (no worthwhile improvement). In epilepsy detection, on the other hand, the patients do not have enough time to react to the seizure. Therefore, there has been much interest in creating tools for seizure forecasting. Epileptic seizure prediction

S. M. Varnosfaderani et al., *Hardware Software Co-design of Epileptic Seizure Prediction Systems*, Analog Circuits and Signal Processing,
https://doi.org/10.1007/978-3-031-95357-6_2

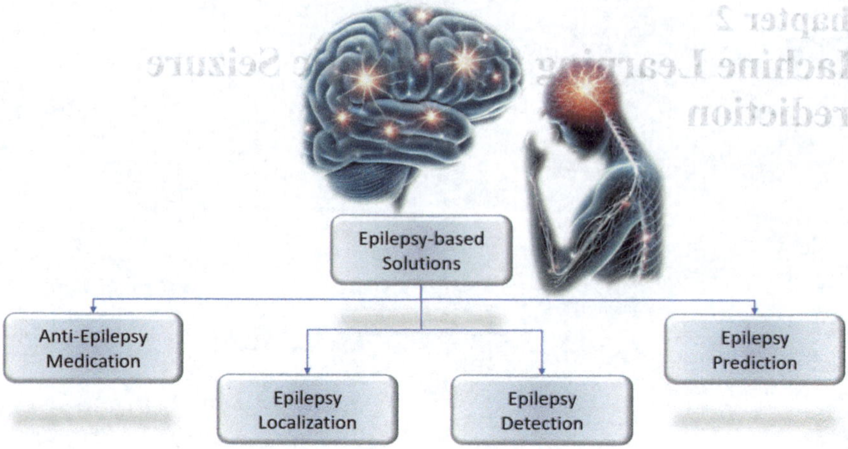

Fig. 2.1 Epilepsy management solutions

has the potential to significantly improve patient care by warning them of imminent seizures and allowing them to prepare a quick-acting drug and take steps to prevent any potential harm. It also paves the way for customized medicines with fewer adverse effects for each individual with epilepsy, and thus seizure intervention systems may be utilized to prevent upcoming seizures [39].

2.2 Epilepsy Prediction

With the aid of machine learning algorithms set up on cloud-based servers, Internet of Things (IoT) technologies have started to play an important role in delivering answers to many health-related issues in recent years. Elderly care [40], remote healthcare [41], exercise programs [42], and the detection and prognosis of neurological and mental diseases including Alzheimer's, epilepsy, autism spectrum disorder, and schizophrenia, among others [43–48], are a few examples of these healthcare solutions. Another paradigm in this area is deep learning [49], which can handle the vast volume of signal data produced by wearable IoT sensing devices like EEG headsets for epilepsy [50]. By requiring less processing time and having the capacity to handle large amounts of multichannel biomedical signal data, algorithms based on deep learning techniques are able to overcome the constraints of classic machine learning algorithms.

Epilepsy detection and prediction systems hold significant promise for effectively managing epileptic seizures. Detection systems identify seizures by distinguishing between the ictal state, when a seizure is occurring, and the interictal state, the period between seizures. This real-time identification allows for timely interventions to help control and manage seizures as they happen. On the other hand, prediction systems aim to anticipate seizures before they occur by differentiating between the interictal phase and the preictal phase, the period leading up to a seizure. By recognizing the preictal phase, these systems can provide early warnings, giving patients crucial time to take protective measures and reduce the risk of injury. Visualization tools further enhance seizure management by making it easier to observe seizure patterns and activity. However, predicting seizures visually remains challenging because subtle differences between preictal and interictal waves are difficult for the human eye to detect. Although automatic seizure detection algorithms have demonstrated good accuracy and performance, they often experience a short delay between seizure onset and detection. Despite this delay, even a brief warning can offer significant protection for patients, allowing them to take immediate action to ensure their safety. Overall, both detection and prediction systems are invaluable in the comprehensive management of epilepsy, offering enhanced safety and improved quality of life for individuals living with this condition.

Early studies on epilepsy prediction were conducted in the 1970s using linear feature extraction methods [51]. In the 1980s, researchers used new nonlinear methods to extract features from EEG signals, taking advantage of their nonlinear properties [52]. The use of the preictal stage for epilepsy identification was also implemented in this decade along with the recognition of the EEG patterns associated with epilepsy, including preictal, ictal, and interictal patterns. Salant et al. conducted early epilepsy prediction roughly 6 seconds before the seizure began in 1998 [53], and Drogenlen et al. expanded on this work in 2003 [54]. They employed a characteristic called Kolmogorov entropy to predict epilepsy 2–40 min before it began. Mormann et al. discovered in 2003 that the phase synchronization of several EEG channels decreases before seizure onset [55] using the theory that the hyper-synchronous firing of neurons in the brain causes epilepsy. Studies based on substantial EEG data have cast doubt on the accuracy of measurements computed in the previous century during the first 10 years of the current century. Researchers discovered that the findings of past research based on a limited number of carefully chosen data could not be replicated on a large amount of previously unreported data.

In worldwide seminars held on the subject, it was determined to hold contests on seizure prediction. These competitions served to standardize the evaluation of the effectiveness of algorithms that had been trained on the same dataset. The International Workshop on Seizure Prediction 3 (IWSP3) and the International Workshop on Seizure Prediction 4 (IWSP4) collaborated on the inaugural seizure prediction competition, which took place in 2007. Three epileptic patients' continuous iEEG recordings were made available to the competitors in both events. The algorithms' performance results, however, fell short of expectations. The 2014 American Epilepsy Society Seizure Prediction Challenge included long-term iEEG recordings of epileptic canines as well as short-term human iEEG including 942 seizures recorded over more than 500 days. The identical training and testing data, each lasting 10 minutes, was given to every contender. An assessment metric for performance was the area under the curve (AUC). Another competition by Melbourne University with the similar format comprised long-term iEEG recording with 1139 seizures [56]. The competitions were open to any algorithm computing the fundamental EEG signal aspects for epilepsy prediction or machine learning techniques trained on these fundamental features. In any case, we are still unsure about the ideal features or algorithms. People entered algorithms that were overly complex in the competitions. Therefore, it is challenging to determine which feature or machine learning (ML) method was better. The contest organizers are currently working with Epilepsyecosystem.org to analyze it. A recent solution presented by Matias Maturana et al. [57] may be effective for a variety of individuals. They determined that the crucial slowing of brain signals served as an epilepsy prediction indication.

2.3 Prior Work on Epilepsy Prediction Systems

Recent advancements in deep learning (DL) techniques have led to the development of highly accurate epilepsy prediction systems [26, 58–67]. To construct an intelligent epilepsy prediction system based on DL, various procedures such as *preprocessing*, *feature extraction*, *classification*, and *postprocessing* are essential. Preprocessing uses filtering, artifact removal techniques, and augmentation to prepare and remove noise from EEG data. In contrast to processing raw EEG signals, the complexity and computational cost of the system are reduced by employing channel-based and segment-based feature extraction [68]. Channel-based feature extraction refers to features extracted from each EEG channel, including statistical time-domain features (such as mean, variance, kurtosis, and skewness), frequency-domain features (such as spectral power in various frequency bands and *short-time Fourier transform (STFT)*), and wavelet-domain features. Systems can have shorter training durations, less complexity, and lower power consumption when handcrafted features are used, compared to systems that automatically extract features using convolutional neural networks (CNNs). In [69], *common spatial pattern* (CSP) is used to extract 324 (18×18) features from 23,040 data points to decrease the number of trainable parameters using CNN; however, the model suffers from low sensitivity. CNN [59, 63–65, 67, 70–72], long short-term memory (LSTM) [60, 73], *bidirectional LSTM* (BiLSTM) [58, 74], and *convolutional gated recurrent neural network* (CGRNN) [62] are examples of classifiers that are used to categorize EEG data. The *K-of-N* [59], mean, and minimum distance (MD) [26] approaches have been utilized as postprocessing techniques to improve classification accuracy, while lowering the false prediction rate (FPR).

A CNN with six convolutional layers and two fully linked layers is developed in [64]. The model is evaluated on two separate sEEG datasets, including a subset of the public CHB-MIT EEG database and the dataset collected at the Mount Sinai Hospital. The CNN obtained a prediction sensitivity of 87.8% and an FPR of 0.142/hr when trained on 3D wavelet tensors generated from the wavelet transformation of raw sEEG data. Similar study can also be found in [75], where STFT is used to convert raw EEG signals into time-frequency characteristics.

The resulting image-like 2D features are then fed into a CNN that includes three convolution blocks and two fully linked layers. When evaluated on datasets from Freiburg Hospital, CHB-MIT, and the American Epilepsy Society (AES) Seizure Prediction dataset, the suggested model has sensitivities of 81.4%, 81.2%, and 75%, respectively. In [76], a CNN model is trained on channel-frequency feature maps, generated by the directed transfer function (DFT), achieving a sensitivity of 90.8% and an FPR of 0.08/h are obtained using Freiburg EEG dataset.

Most of the reported epileptic systems achieved good model performance; however, their high computational resource and memory bandwidth demands restrict their implementation in resource-constrained devices such as wearable and implantable devices. Varnosfaderani et al. [77] proposed a power management model based on patient-specific seizure patterns to control the overall system's power consumption. This model identifies the patient-specific seizure pattern and transitions the system into sleep mode during periods when the likelihood of a seizure is negligible or extremely low. The suggested power management model has the potential to significantly decrease power consumption by 49% compared to a more complex model, with a minimal performance reduction of less than 1%. Recently, LSTMs have been applied in applications for epilepsy prediction due to their capacity for learning long-term dependencies [26, 58, 60, 61]. In [60], LSTM is used to analyze the EEG signals for epileptic seizure prediction, where 643 features in time and frequency domains are selected from each 5-second segment and sent into a two-layer LSTM with 128 memory units in each layer, creating a huge model with more than 500,000 trainable parameters. In [58], 10 channels that obtain the same level of prediction accuracy are chosen based on the highest variance entropy product. The used channel selection method eliminates redundant and unnecessary channels, which brings the total number of trainable parameters down to 18,345 for a deep convolutional autoencoder (DCAE) + BiLSTM model. Although the suggested structure achieves high sensitivity and low FPR, the system's effectiveness is assessed by a small number of patients from CHB-MIT database [78, 79]. In [26], a model with two-layer LSTM showed a good performance with an AUC score of 0.920, verified on the Melbourne dataset. The study uses handcrafted features from the EEG signals, including temporal data (such as mean, variance, and peak-to-peak values) and spectral information (such as spectral power in eight canonical EEG frequency bands), fed into LSTM network. In [80], another LSTM-based seizure prediction technique uses statistical information extracted from EEG rhythms as inputs. When evaluated on the Melbourne dataset, it achieved an AUC score of 0.894.

The imbalanced data problem affects epilepsy prediction systems since there are more interictal samples than preictal samples, thereby resulting in a biased model in favor of interictal samples [81]. A number of methods, such as *overlapping sampling* [59], *generative adversarial network* (GAN) [63], and *deep convolutional generative adversarial networks* (DCGAN) [66], can be used to produce synthetic EEG data to overcome this issue.

2.4 Evaluation Metrics for Epilepsy Prediction

The epileptic seizure prediction systems in the literature are evaluated using four key measures; sensitivity, false prediction rate (FPR), accuracy, and area under the ROC curve (AUC). Sensitivity illustrates how many of the wanted features are detected correctly, and it is the core performance metric, and a high sensitivity is wanted. The false prediction rate illustrates how many unwanted features are detected. For having good results, the number of false positives should be minimal for optimal outcomes [82],[59]. The receiver operator characteristic (ROC) curve is another frequently used assessment tool for the identification and forecasting of ES. An ROC plot is used to demonstrate the diagnostic effectiveness of classifiers. In the interictal and preictal stages, ROC compares the true positive rate against the false prediction rate. The area under the ROC curve (AUC) is also utilized for performance evaluation of ES prediction algorithms [56]. The usage of ROC curves as evaluation metrics is obvious from [83, 84]. Different solutions in the 2016 Kaggle seizure prediction challenge for the Melbourne dataset were ranked using AUC. The following is a description of the most popular assessment measures:

$$\text{Sensitivity} = \frac{\text{TP}}{\text{FN} + \text{TP}}, \tag{2.1}$$

$$\text{FPR} = \frac{\text{TN}}{\text{FP} + \text{TN}}, \tag{2.2}$$

$$\text{Accuracy} = \frac{\text{TP} + \text{TN}}{\text{FN} + \text{TP} + \text{TN} + \text{FP}}, \tag{2.3}$$

where [82]:

- **True positives (TP)**: The number of correct detections of a preictal sample as a preictal
- **False positives (FP)**: The number of incorrect detections of an interictal sample event as a preictal
- **True negatives (TN)**: The number of correct detections of an interictal sample as an interictal
- **False negatives (FN)**: The number of incorrect detections of a preictal sample as an interictal

To evaluate different methods, these definitions are used; however, not all research papers generate any of these four measures explicitly. In general, there is a trade-off between sensitivity and the number of false detections.

2.5 Epileptic Seizure Prediction Models

Machine learning offers a potential solution for precise seizure prediction. Preprocessing, feature extraction, classification, and postprocessing are some of the procedures needed to create an epilepsy prediction system. The several reported prediction techniques are compared in Table 2.1. The research is classified based on the dataset used for making a prediction model. As demonstrated in Table 2.1, to evaluate the models based on Melbourne dataset, the AUC is not reported in some studies. We used four main metrics to report our results. The best results are obtained by CHB-MIT Dataset, while the worst results are reported by models based on the American Epilepsy Society and Melbourne Dataset. There is no explanation for the reason behind this difference. In the following, the mentioned machine learning techniques are compared in terms of several important metrics.

2.5.1 Comparison of Machine Learning Models for Epilepsy Prediction

This section provides a visual and analytical comparison of key machine learning models used in epilepsy prediction. Each model is evaluated based on computational complexity, data requirements, predictive performance, and real-time suitability. These trade-offs are critical for selecting the most appropriate model for specific epilepsy prediction tasks, whether the goal is real-time seizure prediction or high-accuracy offline analysis. Figure 2.2 illustrates the relative positioning of different machine learning models in terms of their computational demands and data requirements. Each model is represented by a marker whose size indicates predictive performance, while the color denotes whether the model is suitable for real-time applications. In this figure, X-axis represents computational complexity. Computational complexity measures the computational resources required for training and prediction. Models with lower complexity (left side of the graph) are easier to implement and run on limited hardware, while those with higher complexity (right side) require more powerful computational resources. Data requirement is shown in Y-axis. This axis reflects how much labeled EEG data is required for the model to achieve effective results. Models that are positioned higher on this axis demand more data, which may be a limitation in clinical settings where large datasets are not always available. In this figure, larger markers correspond to models that achieve higher accuracy and predictive performance in predicting epileptic seizures. These models are typically more complex but offer superior prediction accuracy. Also, green markers indicate models that are more suitable for real-time applications, due to their faster processing times and lower computational overhead. Blue markers represent models that, while highly accurate, are less suitable for real-time prediction due to their computational intensity or data requirements.

Table 2.1 Reported results on epilepsy prediction algorithms

Ref.	Dataset	Preprocessing	Feature extraction	Classification	Postprocessing	Sensitivity %	FPR %	Accuracy %	AUC
[65]	Melbourne	Downsampling	STFT	CNN	–	87.85	–	–	
[66]	Melbourne	Notch filter	STFT	Convolutional epileptic	–	78.11	0.14	–	
	CHB-MIT			Network (CESP)		88.21	0.27		
[67]	Melbourne	Continuous wavelet	–	Semi-dilated convolution	–	88.45	–	–	0.883
	CHB-MIT	Transform (CWT)		Convolution network (SDCN)		98.9	–	–	0.928
[39]	Melbourne	–	EEG scalogram	Multichannel	–	91.15	–	–	0.924
				Vision transform (MViT)					
[85]	Melbourne	–	–	CNN	–	85.2–86.3	–	–	0.914–0.933
[86]	Melbourne	–	–	MLP	–	–	–	–	0.815
[56]	Melbourne	–	Time and Frequency	Random forest, adaptive	–	–	–	–	0.844
[87]	Melbourne	–	Signal energy	Decision tree, kNN	–	33.67	–	–	–
[88]	Melbourne	–	Signal energy,	Logistic regression	–	52.67	–	–	–
			Circadian profile						
[89]	Melbourne	–	EEG spectrogram,	CNN	–	77.36	–	–	–
			Circadian profile						
[26]	Melbourne	Filtering	Statistical features in time domain, power in different frequency bands	Two-layer LSTM	K-of-N, MD, Mean	86.8	0.147	85.1	92
[59]	American Epilepsy Society	–	STFT	CNN	K-of-N method	75	0.21	–	

[60]	CHB-MIT	–	Time/frequency domain, wavelet transform, cross correlation, graph theory	Two-layer LSTM	–	99.28–99.84	0.02–0.11	–	–
[58]	CHB-MIT	–	Deep convolutional autoencoder	BiLSTM	–	99.72	0.004	99.66	–
[73]	CHB-MIT	–	STFT + CNN	LSTM	K-of-N method	98.21	0.13	–	–
[63]	CHB-MIT	Filtering	Combination of common spatial statistics	CNN	Kalman filtering	92.2	0.12	90	–
[62]	CHB-MIT	–	STFT	CRGNN	–	89	–	75.6	
[90]	CHB-MIT	–	Spectral power	SVM		98.7	0.04	–	–
[91]	CHB-MIT	–	Phase locking value	SVM		82.4	–	–	–
[92]	CHB-MIT	–	Mel frequency spectral coefficient	Siamese NN		92.5	–	91.5	–
[93]	CHB-MIT	–	–	CNN		92	0.136	–	–
[94]	CHB-MIT	–	Spectral-temporal feature	Graph convolution Network		95.5	0.109	–	–
[95]	CHB-MIT	–	EEG Spectrogram	LSTM		93	–	–	–
[96]	CHB-MIT	–	EEG Spectrogram	Residual network		89.3	–	92.1	–
[97]	CHB-MIT	–	Mel frequency Spectral coefficients	GNN		94.5	–	95.4	–
[98]	CHB-MIT	–	–	Dilated CNN		93.3	0.007	–	–
[99]	CHB-MIT	–	– Spectrogram	ViT		59.2–97	–	–	–
[74]	Bonn	–	–	Stacked BiLSTM	–	89.21	0.06	91	
[100]	European	CAR-based Method	Time and frequency	Two-layer LSTM	Mean	85.7	0.22	78.1	0.885

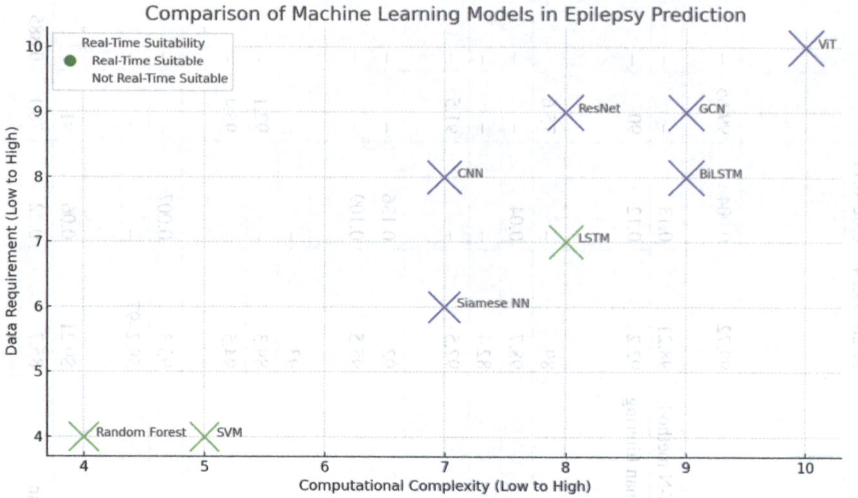

Fig. 2.2 The summery of our contributions

As shown in this figure, support vector machine (SVM) and random forest are models with low complexity and require minimal amounts of data. Both SVM and random forest models require minimal computational resources and can perform well with smaller datasets, making them ideal for real-time epilepsy prediction systems with limited data availability. Their relative simplicity allows for quicker deployment in real-time applications. These models may not capture the full complexity of EEG signals compared to deep learning approaches, resulting in moderate performance. Additionally, they often require manual feature extraction, as they do not natively handle raw EEG data in a temporal context. On the other hand, LSTM networks are specifically designed to handle sequential data, such as EEG signals. Their ability to model temporal dependencies makes them a popular choice for seizure prediction. While they require more computational resources than simpler models, they strike a good balance between performance and real-time feasibility. LSTMs require careful tuning and more data compared to simpler models. However, their overall performance in epilepsy prediction is superior to simpler algorithms like SVM, making them a valuable tool for real-time applications with a moderate computational budget. BiLSTM, ResNet, and graph convolutional network (GCN) models, which are placed in the upper-right quadrant in this figure, are high performers in terms of accuracy due to their ability to capture complex spatial and temporal patterns in EEG data. BiLSTM networks analyze EEG sequences in both directions (past and future), while ResNet and GCN leverage deep learning to detect subtle seizure indicators. The downside of these models is their significant computational complexity and large data requirements. While they provide excellent accuracy, their use is more appropriate for offline analysis or batch processing, rather than in real-time systems. Finally, vision transformers (ViTs) represent the forefront of machine learning technology and are capable of modeling long-range dependencies in EEG signals. This allows them to achieve state-of-the-art performance in epilepsy prediction when sufficient data is available. Since ViT models require large datasets and substantial computational resources to operate effectively, they are not suitable for real-time applications. However, they remain highly valuable in research and diagnostic contexts, where achieving the highest possible accuracy is the primary goal.

2.5.2 Conclusion

In summary, machine learning has profoundly enhanced epilepsy prediction by leveraging a diverse array of algorithms and deep learning techniques to analyze complex EEG data effectively. Traditional models such as support vector machines (SVM) and random forests offer low computational complexity and require minimal data, making them suitable for real-time applications. Meanwhile, advanced architectures like long short-term memory (LSTM) networks, bidirectional LSTMs (BiLSTM), convolutional neural networks (CNNs), and vision transformers (ViTs) provide higher accuracy by capturing intricate temporal and spatial patterns in EEG signals. These machine learning approaches, supported by IoT technologies and cloud-based infrastructures, enable continuous monitoring and early warning systems that can significantly improve patient safety and quality of life. However, challenges persist, including the need for large, high-quality datasets, the high computational demands of sophisticated models, and the difficulty in distinguishing subtle preictal phases from interictal states. Ongoing research, standardized evaluation competitions, and the development of more efficient algorithms are crucial to overcoming these obstacles. Ultimately, the integration of various machine learning techniques holds great promise for creating more accurate, reliable, and accessible epilepsy prediction systems, paving the way for better management and support for individuals living with epilepsy.

Chapter 3
Comprehensive Analysis of EEG Datasets for Epileptic Seizure Prediction

3.1 EEG Signals

Electroencephalography (EEG) signals recording brain activities by electrodes are a helpful source to access the patients' neural activity and play an important role in diagnosis epilepsy. There are two types of EEG signals: *scalp EEG* (sEEG) and *intracranial EEG* (iEEG). sEEG is recorded using electrodes that are placed on the scalp of the subject, as shown in Fig. 3.1a. sEEG is noninvasive and easy to place; however, it cannot be used to record data for a long time. sEEG can be contaminated with different types of artifacts, including motion artifacts. Besides, sEEG uses a low number of electrodes compared to iEEG but covers a larger brain surface.

iEEG signals are recorded using invasive electrodes placed directly on the brain. As illustrated in Fig. 3.1b, these electrodes can either be *subdural* or *depth*. The first is placed on the brain as grids or strips and covers a larger surface area, whereas the latter is inserted deep into the brain, thereby providing higher accuracy [101, 102]. iEEG provides 20 to 100 times higher signal quality than sEEG, is more immune to motion artifacts, and provides seizure localization [103, 104]. The main disadvantage of using invasive electrodes is the risk of complications after the surgery [102].

EEG data can also be classified as *long term* or *short term* based on the duration of the recordings. Short-term EEG recordings are similar to routine EEG scans, whereas long-term EEG recordings span for extended periods, from months to years. Long-term EEG is usually performed to correlate clinical behavior with an EEG phenomenon. EEG recordings may contain unusable data, called *dropout*, as shown in Fig. 3.2. They are caused by a machine-related error and should be removed for better processing. Dropout could be a constant value, a constant pattern, or low amplitude values and may span from a few seconds to a couple of hours. Small intervals of dropout data can be ignored.

S. M. Varnosfaderani et al., *Hardware Software Co-design of Epileptic Seizure Prediction Systems*, Analog Circuits and Signal Processing,
https://doi.org/10.1007/978-3-031-95357-6_3

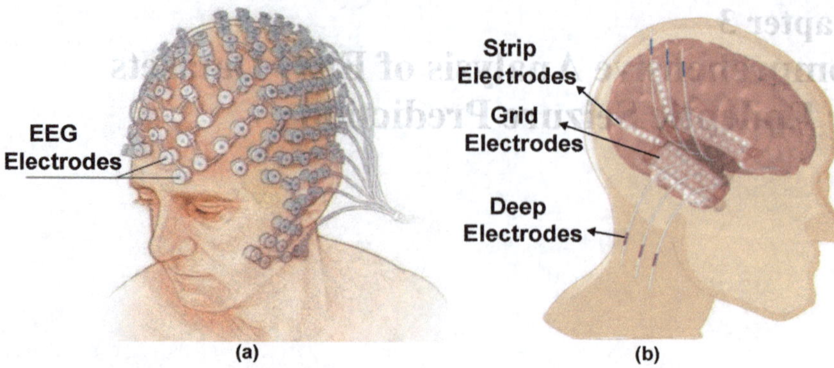

Fig. 3.1 Placement of electrodes in (**a**) sEEG (Source: *Mayo Foundation*) and (**b**) iEEG (Source: *Epilepsia ILAE*)

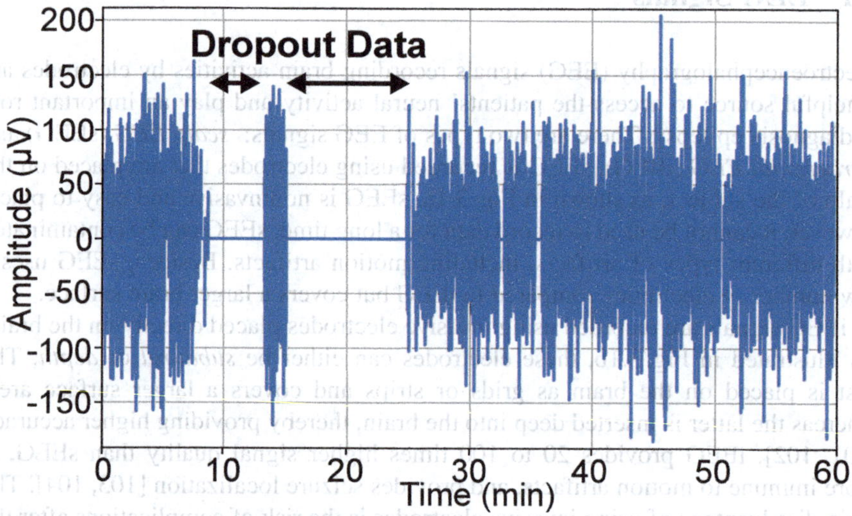

Fig. 3.2 An example of dropout data from an epileptic patient

3.2 Discussion and Comparative Analysis of Various Datasets

Public datasets are available online, and they can be accessed free of cost for noncommercial uses. In this part, the details of the Melbourne and American Epilepsy Society datasets which are iEEG signals and CHB-MIT and Bonn datasets including sEEG are discussed.

3.2.1 Melbourne University Seizure Prediction Challenge Data

Melbourne University provides an EEG dataset that contains long-term iEEG, recorded from three patients with the lowest seizure prediction performance from the NeuroVista trial [56]. Invasive strip (4–4) electrodes are placed on the focal hemisphere of the patients to record a total duration of 1,155 hours of data with 16 channels and a sampling rate of 400 Hz. The data was used in a seizure prediction competition held on Kaggle and then was provided as part of a crowd-sourcing ecosystem for epilepsy prediction. The dataset contains one hour of preictal recording before each seizure, which is broken into six 10-minute files. There are also 5-minute gaps between preictal recording and seizure occurrence, called seizure horizon onset, which allows enough time to predict and warn the patient for an oncoming seizure. Moreover, to prevent overlapping with postictal and preictal stages, interictal segments are recorded with a gap of 4 hours before and after each seizure. The duration of interictal segments is one hour, which is split into six 10-minute files [56]. The data is divided into training and testing, which are placed in two different folders.

One of the main challenges in using this dataset for epilepsy prediction is that the recording of data took more than one year. The testing data was recorded 6 months after the training data in which the patient was under medication, thereby causing significant differences between the distribution of the test and train sets and complicating the prediction task This causes a common problem named *covariate shift*, *concept drift*, or *data mismatch* [56]. The Melbourne dataset is in .mat format and can be easily read using MATLAB or the SciPy library of Python. Table 3.1 shows the details of the dataset, including recording duration, the number of seizures, the number of channels, and sampling rate.

Table 3.1 Details of the Melbourne dataset

Dataset features	Patient 1	Patient 2	Patient 3
Gender	Female	Female	Female
Age (years)	22	51	50
Recording period (days)	559	393	374
Total recording (hours)	173	507	475
# of seizures	390	204	545
# of train files (interictal)	570	1836	1908
# of train files (preictal)	256	222	255
# of test files (interictal)	16	18	18
# of test files (preictal)	46	279	188
# of channels	16	16	16
Sampling rate (Hz)	400	400	400

Table 3.2 Details of the American Epilepsy dataset

Case	# of Channels	Sampling rate (Hz)	# of train files	# of test files	Duration (hours)
Patient 1	15	5000	68	195	43.8
Patient 2	24	5000	60	150	35
Dog 1	16	400	504	502	167.7
Dog 2	16	400	542	1000	257
Dog 3	16	400	1512	907	403.2
Dog 4	16	400	901	990	315.2
Dog 5	15	400	480	191	111.8

3.2.2 American Epilepsy Society Dataset

The American Epilepsy Society dataset contains iEEG data of seven subjects (two humans and five dogs) with a total duration of 1,333.7 hours. The number of channels varies from 15 to 24 among the subjects, while the sampling rates are 5000 Hz and 400 Hz for humans and dogs, respectively. The recordings are structured similar to the Melbourne dataset with 1 hour of recordings divided into six 10-minute files and a seizure horizon onset of five minutes. Interictal files from the dogs have a week gap with seizures, while interictal files from humans have a gap of 4 hours [105].

Several hours of EEG data for the first human subject contain dropout data across all channels, caused by 60 Hz noise, which is coupled through recording machines. Notch filters could be used to remove this noise. The details of the dataset are shown in Table 3.2.

3.2.3 CHB-MIT sEEG Dataset

The CHB-MIT dataset contains sEEG data recorded at a sampling rate of 256 Hz from 23 epileptic patients at the Children's Hospital of Boston. The dataset is divided into cases, each representing one patient, except for case 1 and case 21, where the data are recorded from the same patient with a gap of 1.5 years. Although the number of channels varies from 23 to 26, the following 18 channels are common among all patients: FP1-F7, F7-T7, T7-P7, P7-O1, FP1-F3, F3-C3, C3-P3, P3-O1, FZ-CZ, CZ-PZ, FP2-F4, F4-C4, C4-P4, P4-O2, FP2-F8, F8-T8, T8-P8, and P8-O2. Unlike Melbourne and American Epilepsy Society datasets, the CHB-MIT dataset does not contain preictal or interictal labels but can be extracted using seizure timings provided in meta-data files of every patient. This can be applied for all cases except case 24, where the start and end times of the file are not specified. The dataset has 198 seizures, and the files are in .edf format [78]. The details of this dataset are shown in Table 3.3.

Table 3.3 Details of the CHB-MIT dataset

Case	Gender	Age (years)	# of seizures	Duration of recordings (hh:mm:ss)
chb01	Female	11	7	40:33:08
chb02	Male	11	3	35:15:59
chb03	Female	14	7	38:00:06
chb04	Male	22	4	156:03:54
chb05	Female	7	5	39:00:10
chb06	Female	1.5	10	66:44:06
chb07	Female	14.5	3	67:03:08
chb08	Male	3.5	5	20:00:23
chb09	Female	10	4	67:52:18
chb10	Male	3	7	50:01:24
chb11	Female	12	3	34:47:37
chb12	Female	2	40	20:41:40
chb13	Female	3	12	33:00:00
chb14	Female	9	8	26:00:00
chb15	Male	16	20	40:00:36
chb16	Female	7	10	19:00:00
chb17	Female	12	3	21:00:24
chb18	Female	18	6	35:38:05
chb19	Female	19	3	29:55:46
chb20	Female	6	8	27:36:06
chb21	Female	13	4	32:49:49
chb22	Female	9	3	31:00:11
chb23	Female	6	7	26:33:30
chb24	Not provided	Not provided	16	21:17:47

3.2.4 Bonn Dataset

The Bonn EEG dataset contains sEEG and iEEG from five subjects with a sampling rate of 173.61 Hz. The data is collected from patients experiencing a seizure-free interval (folders N and F), from patients having a seizure (folder S), and from healthy patients at rest (folders Z and O). There are 100 samples in each folder where every sample has a record duration of 23.6 seconds. Unlike other available datasets, the Bonn dataset contains healthy labels. However, due to the short total duration of 3.3 hours and the use of a single channel to record the data, the Bonn dataset is not preferred for epilepsy prediction algorithm development [107].

3.2.5 *The European Epilepsy Dataset (Private)*

The European Epilepsy dataset is the largest available EEG dataset, including data for 30 invasive patients over multiple days of recording and a total duration of 6,488 hours. Although there are two packages, scalp and intracranial EEG, this chapter only refers to the intracranial EEG package of the European Epilepsy dataset. The number of channels and sampling rates vary widely among the patients ranging from 31 to 124 and 256 Hz to 2500 Hz, respectively, which are much higher than other available datasets. The dataset includes a few scalp electrodes for electrooculography (EOG) and electrocardiogram (ECG), which record the electrical activity of the eyes and heart, respectively. The files of this dataset are not labeled as preictal or interictal; however, they can be extracted using the timing information of the 590 seizures present in the dataset. Further, the dataset provides information about seizure types and origin of the seizure, which can help develop patient-specific machine learning algorithms to predict the seizure location in the brain. Finally, the dataset includes details about subclinical seizures, which show abnormal electrical activity in the brain with no physical symptoms. The details of the dataset are shown in Table 3.4.

3.3 Summary and Overall Comparison

Table 3.5 summarizes and compares all the datasets, where the European Epilepsy dataset provides the largest number of subjects, the number of channels, and duration. This makes it the most comprehensive dataset among those listed, offering significant versatility for researchers developing epilepsy prediction models. With 30 subjects, between 31 and 125 channels, and a total recording duration of 6,488 hours, it enables robust model training and provides extensive data for both ictal and nonictal events. The wide range of sampling rates (256–2500 Hz) further enhances its applicability for analyzing various types of seizure patterns with high granularity. The availability of ictal labels ensures that supervised learning methods can effectively be applied, making it a highly valuable resource for seizure prediction.

In contrast, the CHB-MIT dataset, while also substantial in terms of subjects (23) and the number of channels (23–26), lacks ictal labels, which poses a challenge for models that require well-labeled seizure events. The sampling rate of 256 Hz is sufficient for most EEG analyses, but the dataset has a shorter duration (979 hours), which limits its ability to capture longer-term patterns. The American Epilepsy Society (AES) dataset is another valuable resource with a smaller number of subjects (7) but offers a high sampling rate range (400–5000 Hz), which is beneficial for high-resolution seizure detection. The Melbourne dataset, although it contains

Table 3.4 Details of the European Epilepsy dataset

Patient	Gender	Age	Samp. rate (Hz)	# of channels	# of seizures	Duration (hours)
FR_264	F	48	256	67,91	8	140.5
FR_253	F	37	512	31,58	7	262.1
FR_1146	M	22	1024	116	26	113.6
FR_1096	F	32	1024	100	9	163
FR_1084	F	48	1024	93,117	94	250.6
FR_970	M	15	256	119,122	19	205.1
FR_958	F	14	1024	97,121	16	232.2
FR_922	M	39	1024	85,109	30	114.6
FR_916	M	23	1024	85,109	52	430.8
FR_862	F	16	256	61,85	9	233.6
FR_620	M	42	1024	39,63	7	255.1
FR_590	M	18	1024	99,122	13	252.2
FR_583	F	22	256,1024	41,65	23	203.5
FR_565	M	13	256	93,117	14	229.1
FR_548	M	17	1024	85,109	31	143.6
FR_384	F	50	256,1024	95,119	15	139
FR_375	M	35	256	57,81	26	184
FR_273	F	3	256	47,68	7	203.4
FR_1150	F	32	1024	125	9	158.7
FR_442	M	21	1024	71,95	22	186
FR_139	F	53	256,512	45,60	6	172.6
FR_115	M	34	256	86	26	229.4
FR_1125	F	11	1024	63,87	14	158.5
FR_1077	F	29	1024	122	9	185.8
FR_1073	F	47	1024	72,96	20	228.8
FR_818	F	27	1024	47,71	9	252.7
FR_635	F	63	1024	82	21	119.2
FR_CM_32	M	10	2500	102	9	207.1
PA_inv_13245	F	43	1024	57	31	357.6
PA_inv_13089	F	29	1024	64	8	475.9

only three subjects and 16 channels, provides a significant duration (1,155 hours) and ictal labels, making it suitable for seizure event detection over long periods. Lastly, the University of Bonn dataset, while useful for basic comparisons, is the most limited with only five subjects, one channel, and a short duration of 3.3 hours, which constrains its utility in large-scale studies.

Table 3.5 Summary of the available EEG datasets

Dataset	Type	Ictal label	# of subjects	# of channels	Sampling rate (Hz)	Duration (hours)
Melbourne	iEEG	Yes	3	16	400	1,155
American Epilepsy Society	iEEG	Yes	7	15–24	400–5000	1,333.7
CHB-MIT	sEEG	No	23	23–26	256	979
University of Bonn	sEEG + iEEG	Yes	5	1	173.61	3.3
European Epilepsy dataset	iEEG	No	30	31–125	256–2500	6,488

3.4 Dataset Usage in Epilepsy Prediction Systems: Discussion, Challenges, and Insights

Two main metrics for epilepsy prediction are generally used to evaluate the prediction accuracy of machine learning models, namely, *sensitivity* and *false prediction rate* (FPR). Sensitivity is defined as the probability of seizures to be correctly predicted, while FPR is the rate of detecting seizures incorrectly per hour. *Accuracy* can be derived from sensitivity and FPR [82, 108]. Other metrics include *complexity* and *power consumption*, which are of great importance, especially in hardware accelerators.

An epilepsy prediction system requires various steps, including *preprocessing*, *feature extraction*, *classification*, and *postprocessing*. Preprocessing aims to prepare and clean the data using filtering, artifact removal methods, and augmentation. Feature extraction can be used to extract the most important features in the *time domain* such as mean, variance, kurtosis, and skewness, in the *frequency domain* such as spectral power in different frequency bands and *short-time Fourier transform (STFT)*, and in the time-frequency domain such as *wavelet transform* [109]. Various classifiers are used to classify EEG data, including *convolutional neural networks* (CNN) [59, 63, 65], *long short-term memory* (LSTM) [73, 106], *bidirectional LSTM* (BiLSTM) [58, 74], and *convolutional gated recurrent neural network* (CGRNN) [62]. Postprocessing methods may be used to enhance classification accuracy, using the *K-of-N* method [59], which also reduces the false prediction rate (FPR). Schemes that use handcrafted features instead of automatic feature extractions (by CNN as an example) can have lower training times, complexity, and power consumption.

Table 3.6 compares various reported prediction schemes. The performance depends not only on the machine learning algorithm and the used methods for preprocessing and postprocessing but also on the used dataset. The worst results, reported in [65] and [59], are for the Melbourne and American Epilepsy Society datasets, respectively. Both of these datasets have data drift, where EEG data is collected at different times (6 months apart in Melbourne), and thus are hard to train compared to other datasets. A possible solution is to extract non-time-domain features from the data, such as frequency-domain features [65].

Due to the limitation in recording the data from patients, a low number of samples is another challenge in datasets, which might lead the model to biased machine learning performance. To increase the number of samples, different datasets can be combined while ensuring that the electrode locations match. For example, in [110], the Bonn dataset is used to train the system, and the CHB-MIT dataset is used to validate and test the system.

Table 3.6 Reported results on epilepsy prediction algorithms

Ref.	Dataset	Preprocessing	Feature extraction	Classification	Postprocessing	Sensitivity %	FPR %	Accuracy %
[65]	Melbourne	Downsampling	STFT	CNN	–	87.85	–	–
[59]	American Epilepsy Society	–	STFT	CNN	K-of-N method	75	0.21	–
[106]	CHB-MIT	–	Time/frequency domain, wavelet transform, cross correlation, graph theory	Two-layer LSTM	–	99.28–99.84	0.02-0.11	–
[58]	CHB-MIT	–	Deep convolutional autoencoder	BiLSTM	–	99.72	0.004	99.66
[73]	CHB-MIT	–	STFT + CNN	LSTM	K-of-N method	98.21	0.13	–
[63]	CHB-MIT	Filtering	Combination of common spatial statistics	CNN	Kalman filtering	92.2	0.12	90
[74]	Bonn	–	–	Stacked BiLSTM	–	89.21	0.06	91
[62]	CHB-MIT	–	STFT	CRGNN	–	89	–	75.6

3.5 Conclusion

In this chapter, we analyzed and compared various EEG datasets used for epileptic seizure prediction. The datasets were compared in terms of the number of patients, the number of channels, types of electrodes, sampling, rate, and recording period. The European Epilepsy dataset contains the longest recording duration with the highest number of patients compared to other datasets, and thereby can be an excellent candidate to develop accurate machine learning models to predict epilepsy. Various machine learning models for epilepsy prediction were analyzed and compared, and the impact of the dataset was discussed. This work can be of great importance for researchers in EEG prediction systems.

Chapter 4
Comprehensive Analysis of EEG Datasets for Epileptic Seizure Prediction

4.1 Analysis of Artifacts Removal Techniques in EEG Signals for Energy-Constrained Devices

Electroencephalography (EEG) is an electrophysiological monitoring method, which records the neural activities of the brain. EEG is a useful tool to diagnose various brain disorders, including epilepsy and Parkinson's diseases [111]. Many commercial products, including wearables, are used to acquire EEG signals [112, 113], but signal quality could be deteriorated by undesired artifacts, which can be grouped into two main categories: *extrinsic* and *intrinsic* [114]. The extrinsic artifacts include environmental noise and instrumental error, such as faulty electrodes and line noise [115]. This type of noise can be mitigated by applying precise recording procedures. The intrinsic artifacts are caused by human activities, including ocular, cardiac, and motion artifacts [115–117].

The topic of artifact removal is an open research problem, where the denoising technique should be selected based on the application and signal type [115]. For example, scalp EEG (sEEG) is mainly affected by motion artifacts [118], while intracranial EEG (iEEG) is mostly sensitive to ocular and cardiac artifacts [103]. Figure 4.1 shows the frequency ranges of the EEG signals. As depicted in the figure, ocular and motion artifacts are the most challenging to remove since their frequency band lies within the EEG frequency band. In particular, ocular artifacts tend to have high-amplitude waveform, making it easier to detect and remove [115].

Various techniques are reported in the literature to remove artifacts and noise from EEG, as shown in Table 4.1, including wavelet transform (WT), blind source separation (BSS), empirical mode decomposition (EMD), and moving average filter [107, 119–127]. Each technique is suitable for a particular application based on the given requirements. For example, BSS effectively separates the noise from the signal, but it requires high computational power. EMD is a data-driven approach that is sensitive to noise. Moving average filter removes the noise based on a

S. M. Varnosfaderani et al., *Hardware Software Co-design of Epileptic Seizure Prediction Systems*, Analog Circuits and Signal Processing,
https://doi.org/10.1007/978-3-031-95357-6_4

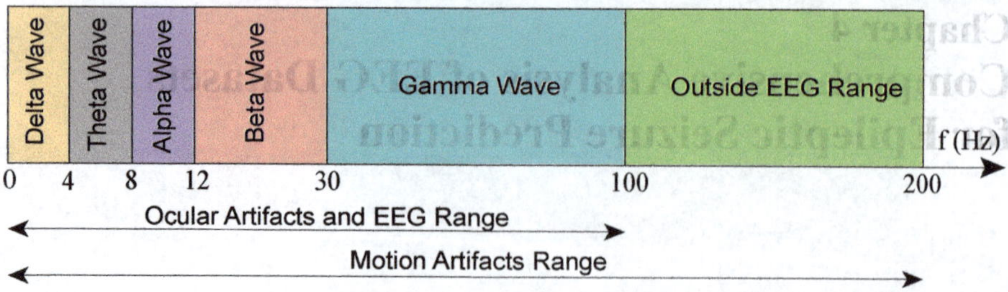

Fig. 4.1 The frequency range for EEG signals and various artifacts

Table 4.1 Comparisons

Method name	Pros.	Cons.
Wavelet transform [119]	• Low computational complexity • Highly customizable • Capable of real-time processing	• Signal could be removed if overlaps with noise
Blind source separation [120]	• Can separate signal and noise	• Not efficient on single-channel EEGs • High computational complexity • Not capable of real-time processing
Empirical mode decomposition [121]	• Data-driven technique • Low computational complexity • Capable of real-time processing	• Sensitive to noise
Moving average filter [122]	• Low computational complexity • Capable of real-time processing	• No effective in removing certain types of noise

time-domain filter; however, it is not effective in removing certain noise types such as motion artifacts. WT promises a real-time, low-computational, and highly customizable method that is more suited for energy-constrained applications such as wearable devices.

4.1.1 Background Information and Related Work

4.1.1.1 Datasets

Two databases are used: Motion-Artifact Contaminated EEG [125] and Bonn databases [107]. The first is an sEEG, collected from six different healthy subjects consisting of four experiments for each one, in 24 trials, and a sampling rate of 2048 Hz. Two electrodes are placed 20 mm apart, where one electrode is used to record the sEEG with a 1-minute motion artifact in every 2 minutes of recording, and the second electrode is used to record the ground truth sEEG [125]. The Bonn database consists of five folders of single-channel sEEG and iEEG segments, where each folder consists of 100 samples recorded at 173.61 Hz for 23.6-second segments. Folders Z and O include sEEG segments, which do not contain artifacts, while folders F, N, and S consist of iEEG segments that are mainly contaminated with ocular artifacts. To evaluate the performance of different methods in removing ocular artifacts, folders F, N, and S are selected.

4.1.1.2 Denoising Techniques

Wavelet Transform WT is used to decompose a time-domain signal into the time-frequency domain with varying levels of accuracy in the frequency domain [119]. The components of the original signal obtained from the WT are known as wavelet coefficients. The original signal can be reconstructed by a linear combination of the extracted Wavelet coefficients, which can be extracted based on a mother Wavelet and a desired level of decomposition. The selection of the mother Wavelet is related to the shape of the signal being analyzed. In this chapter, the Daubechies-4 mother wavelet is selected due to its similar shape to interictal spikes in EEG signals [128].

The selection of the decomposition level depends on how fine-scaling the analysis must be and how much computational complexity can be afforded. In this chapter, the signal is decomposed into 7 and 11 levels of decomposition for Bonn and Motion-Artifact Contaminated EEG databases, respectively. The WT can be defined as the inner product of the signal X_n with the scaled and time-shifted wavelet $\psi_{x_n}[m]$ [4]. In particular, the WT is given by

$$W_{x_n}(a, b) = < x_n, \psi_{a,b} >, \tag{4.1}$$

where a and b are the scale and shift parameters, respectively, x_n is the input signal, and ψ is the scaled and shifted version of the mother Wavelet.

Two broad classes of WT can be used for denoising signals: *WT with autocorrelation threshold* and *WT with shrinkage rules*. Once this set of wavelet coefficients is obtained, the WT with autocorrelation threshold method calculates the autocorrelation at lag level 1 for each coefficient. If the autocorrelation of any coefficient is greater than a specific value, the entire coefficient is removed from the set of wavelet coefficients before creating the denoised output signal

[124]. WT with shrinkage rules denoises a given signal by reducing the values in the signal on a sample-to-sample basis. Specifically, the important part of the EEG is preserved, while the coefficients for the other parts are simply thrown away. Shrinkage rules must be coupled with a thresholding rule to decide on the samples with the value below, above, or at a specified threshold value. The shrinkage rules can be *soft thresholding* or *hard thresholding*, where both reduce samples whose absolute values fall below or reach the threshold value to 0; however, they differ in handling values above the threshold. Hard thresholding keeps these values unchanged, whereas soft thresholding shrinks the values toward zero.

One of the following three alternatives algorithms of WT with shrinkage rules can be used:

- *SURE Shrink Algorithm*: The extracted wavelet coefficients are denoised according to the threshold value calculated by running an algorithm based on *Stein's unbiased risk estimate* (SURE) [117]. This threshold is obtained by calculating Stein's unbiased estimate of risk for each sample in the signal and using the minimum value calculated as the thresholding value.
- *Universal Threshold* (UT): The threshold value, K, is determined by

$$K = \sqrt{2logN} \times \sigma, \tag{4.2}$$

$$\sigma^2 = \text{median}\left(\frac{|C_a|}{0.6745}\right), \ \forall a = 1, 2, \ldots, j, \tag{4.3}$$

  where σ is the standard deviation, σ^2 is the variance, N is the number of data to be processed, $|C_a|$ is the absolute values of the wavelet coefficients vector for the j levels of decomposition, and 0.6745 is the constant value for Gaussian noise [126].
- *Statistical Threshold* (ST): The threshold, T, for the wavelet coefficients vector is given by

$$T = 1.5 \times \text{std}(C_a), \ \forall a = 1, 2, \ldots, j, \tag{4.4}$$

  where $std(C_a)$ is the standard deviation of the Wavelet coefficient vector for the j levels of decomposition [126].

Moving Average Filter A moving average filter is used in this work to remove the noise from the EEG databases. The implemented moving average filter uses a cascaded moving average algorithm to create a finite-impulse response filter (FIR). The filter has two stages: an I-point moving average overlapping with a J-point moving average filter [122].

4.1.2 Evaluation Methodology, Results, and Analysis

This section presents the metrics used to investigate the denoising methods and simulation results in MATLAB for artifacts removal techniques on the EEG databases.

4.1.2.1 Efficacy Metrics

We compare the denoising techniques in terms of two major metrics: SNR and NMSE. SNR can be expressed as a ratio of the power of the signal to the power of the noise, which can be defined by

$$\mathrm{SNR} = 10\log_{10}\frac{\sigma_{\hat{x}}^2}{\sigma_e^2}, \tag{4.5}$$

where $\sigma_{\hat{x}}^2$ is the variance of the denoised signal and σ_e^2 is the variance of the error signal. A positive SNR indicates that the power of the signal is larger than the power of the noise, which suggests an effective denoising method. On the other hand, a negative SNR indicates larger noise than the signal, which suggests a poor performance of the denoising technique. By contrast, NMSE is defined by

$$\mathrm{NMSE} = 10\log_{10}\frac{\Sigma_{m=1}^{M}\left(x_n[m]-\hat{x}_n[m]\right)^2}{\Sigma_{m=1}^{M}\left(x_n[m]\right)^2}, \tag{4.6}$$

where $x_n[m]$ is the original contaminated EEG signal, and $\hat{x}_n[m]$ is the denoised EEG signal [117, 122]. As indicated in Eq. 4.6, the smaller the error, $(x_n[m] - \hat{x}_n[m])$, the larger the negative values of NMSE in decibels, and thus better performance of the denoising techniques.

4.1.2.2 Overall Results Under the Bonn Database

Figure 4.2 compares the moving average method (as implemented in [127]), autocorrelation method (as implemented in [125]), and the three shrinkage rules, which include SURE Shrink, universal thresholding (UT), and statistical threshold (ST) for the Bonn database in terms of SNR and NMSE. Figure 4.2a shows average NMSE and SNR for each folder (F, N, S), and Fig. 4.2b averages NMSE and SNR for all the folders. All shrinkage rules methods here employ hard thresholding. Note that this thresholding pertains only to the WT with shrinkage rule methods since the moving average filter and WT with autocorrelation methods do not make use of a thresholding rule. As shown in Fig. 4.2b, the WT with SURE Shrink using hard thresholding performs the best in both average SNR and NMSE. The SURE Shrink method reduces the high amplitude of the ocular artifact while preserving

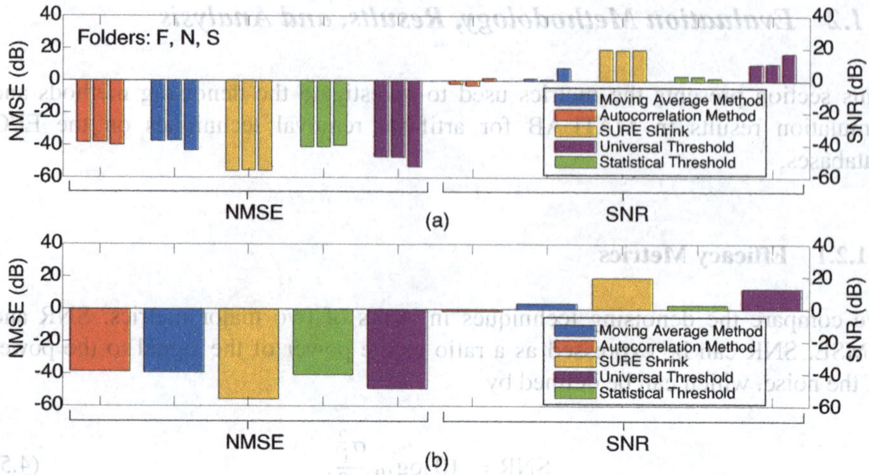

Fig. 4.2 Comparing various methods over all trials of the Bonn EEG database (hard thresholding when applicable). (**a**) Average NMSE and SNR for each folder. (**b**) Average NMSE and SNR for all the folders

the original EEG signal. The results agree with the reported work in the literature when using SURE Shrink [129] and WT denoising methods [115, 126, 129]. The WT with autocorrelation performs the worst in terms of average SNR, due to the nature of the method, which caused removing the original signal in the denoising process.

4.1.2.3 Overall Results Under the Motion-Artifact Contaminated Database

Figure 4.3 shows the comparative results under the Motion-Artifact Contaminated database in terms of SNR and NMSE. Once again, all shrinkage rules methods here employ hard thresholding. As depicted in Fig. 4.3, the three shrinkage rule methods perform better than other techniques due to their data-driven specific approach to denoising. They remove more specific pieces of noise than other methods due to their sample-to-sample process, thus maintaining the original signal. The universal threshold using hard thresholding performs the best since it takes into account the variance of the original signal, where motion artifacts have a large amount of variance in terms of amplitude and frequency range. The moving average filter and WT with autocorrelation threshold perform similarly for the worst denoising methods. This is partially due to the moving average filter not removing enough noise and primarily focusing on simply detrending the signal in time domain. WT with autocorrelation performs worse than others primarily due to the lack of specificity in removing the noise, i.e., it removes parts of the original signal.

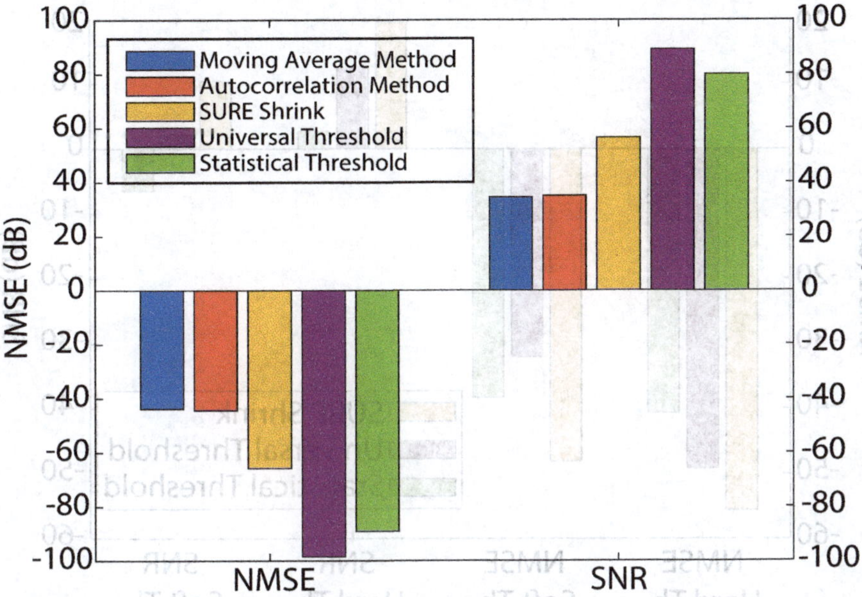

Fig. 4.3 Comparing various methods over all trials of the Motion-Artifact Contaminated database (hard thresholding when applicable)

4.1.2.4 Analysis of Shrinkage Rules

Let us now compare the effectiveness of shrinkage rules methods and analyze the impact of thresholding, hard versus soft thresholding. As shown in Figs. 4.4 and 4.5, the use of the soft thresholding tends to decrease average SNR and increase average NMSE after denoising for each method when tested on both databases. This is due to soft thresholding reducing parts of the signal that fall above the noise threshold, whereas the hard thresholding rule leaves them and only removes or reduces values that fall below the threshold value. In particular, the average SNR obtained by the statistical threshold using soft thresholding has negative values in the Bonn database, suggesting that the method does not remove the noise but rather removes the original signal.

4.1.3 Discussion on Computational Complexity of Denoising Techniques

WT has lower complexity and power consumption compared to other techniques and can achieve real-time operation with low cost, which is useful for energy-constrained applications [115, 117]. The most expensive method is the WT with SURE Shrink due to the complexity of the SURE algorithm [117], while the

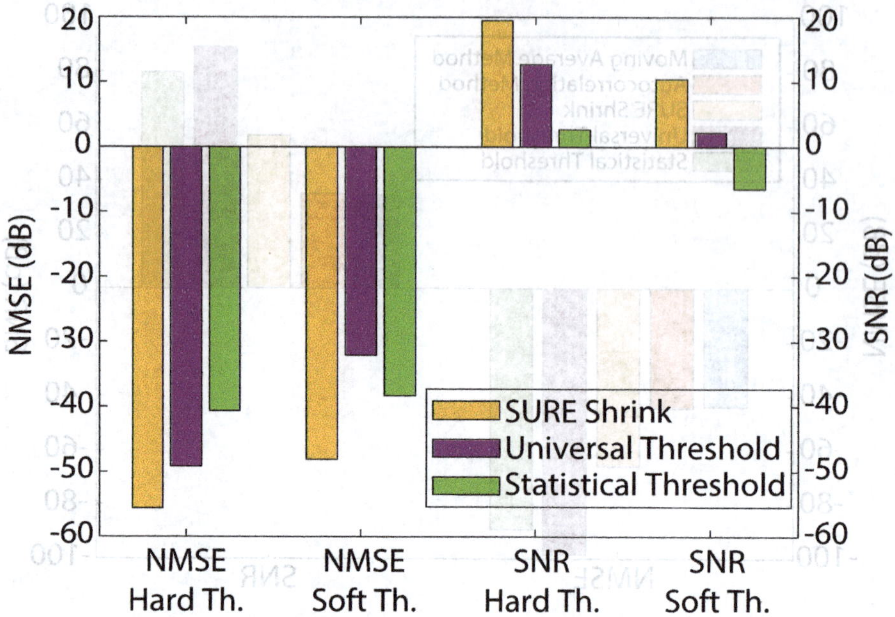

Fig. 4.4 Comparing shrinkage rules methods over all trials of all the folders of the Bonn EEG database (hard and soft thresholding)

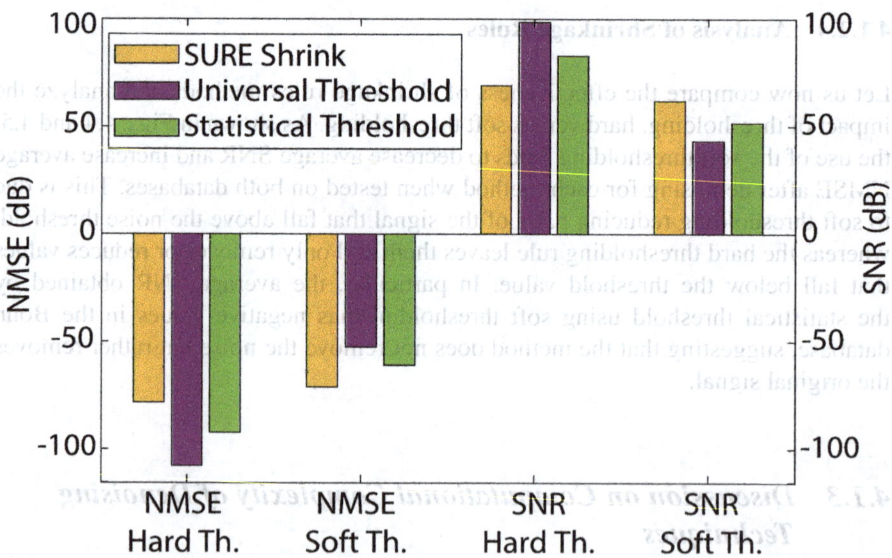

Fig. 4.5 Comparing shrinkage rules methods over all trials of the Motion-Artifact Contaminated database (hard and soft thresholding)

WT with autocorrelation has the least computational complexity. In addition, WT with the universal threshold and WT with the statistical threshold require similar computational power due to relatively simple implementation and the similarities in how both threshold values are computed. Overall, for WT-based methods, the limiting factor in computational complexity is the wavelet transform itself due to the complexity of some mother wavelets, such as the Daubechies family, and higher decomposition levels. To reduce the computational complexity and the power consumption of WT-based methods, a lower level of decomposition can be used, such as level 5 or level 6, with a simple mother wavelet such as the Haar mother wavelet. The moving average filter method has variable computational complexity due to the variable window size, but the power consumption should not vary significantly.

4.2 Artifact Removal Based on Common Average Reference

The detailed results and analysis are of great importance to researchers and practitioners in EEG systems and to achieve real-time, low-power, and low-cost denoising methods for energy-constrained devices.

Re-referencing techniques are used to remove the common noise that is coupled to EEG electrodes due to the moving of the reference electrode or the noise coupled from the data acquisition instrument. There are several re-referencing techniques that can be used to reduce the effect of the common noise, including bipolar, Laplacian, and average methods. The average or Laplacian technique is commonly used for subdural electrodes, while the bipolar or Laplacian method can be used for stereo EEG (SEEG) electrodes [130–132].

While iEEG is less affected by noise compared to scalp EEG, a scalp or epidural reference electrode is commonly used to record the iEEG data, and thus the noise caused by the scalp electrode can be coupled to intracranial electrodes [133]. Therefore, re-referencing techniques are necessary to remove noise from the scalp electrode as well as other common artifacts.

There are many re-referencing techniques that are reported in the literature as follows. Huang et al.[132] compared the performance of five re-referencing methods: common average reference (CAR), bipolar reference, electrode shaft reference (ESR), Gray-white matter reference (GWR), and Laplacian reference, where the Laplacian re-referencing method is a good option for broadband gamma and low-frequency oscillatory activity. In [131], the common noise is successfully eliminated using the bipolar reference and CAR for depth electrodes and grid electrodes, respectively. In [134], an automatic scalp reference removal from iEEG data is achieved using independent component analysis (ICA). The results are compared and evaluated with bipolar and CAR methods using iEEG recordings from three patients for epilepsy surgery purposes.

Table 6.1 summarizes the reported methods for removing the common noise from EEG data. The most commonly used methods for removing common noise are bipolar reference, which involves subtracting each electrode from one or more of its neighbors, and CAR, which involves subtracting each electrode from the average of all electrodes. Note that due to the close proximity of the iEEG electrodes, there is a great probability of data loss when using these two methods.

Since ICA is a signal processing method used to separate independent sources, it can also be used to remove common noise. ICA decomposes the signal into several independent components, equal to the number of electrodes, where the reference-related component needs to be taken out. After that, the signal can be reconstructed using the decomposed components. ICA requires a large number of electrodes to obtain good results, causing significant overhead. Further, finding the component that is related to the reference electrode is not a straightforward task.

We present a detailed preprocessing technique for removing various artifacts from iEEG data, including a patient-specific re-referencing method. This technique is evaluated using the European iEEG dataset to assess its effectiveness [100]. In the European Dataset, since the scalp EEG is utilized as a reference in this dataset, this method reduces the common noise caused by the scalp reference, which is one of the main noise sources in this dataset. Additionally, in this chapter, we offer in-depth evaluations of the impact of various noise types on iEEG data in the frequency and time domains.

4.2.1 The European iEEG Dataset and Its Challenges

The European Epilepsy dataset is the largest available collection of long-term EEG recordings. This dataset consists of multiple-day recordings from 30 patients with a total duration of 6,488 hours [136, 137]. The sampling rate and the number of channels vary among the patients from 256 Hz to 2500 Hz and 31 to 124, respectively. The labels of this dataset are not determined; however, they can be extracted based on the seizure onset time provided for 590 seizures. Further, information regarding the electrode schemes and the locations of the seizure onset zone electrodes is provided.

The dataset is made up entirely of raw data without any preprocessing, causing challenges in cleaning the data from various noise sources. The dataset includes iEEG which is less susceptible to noise; however, a scalp EEG electrode is utilized as a reference electrode. Therefore, the noise caused by the scalp electrodes is coupled to other electrodes and considered as a common noise. Figure 4.8 highlights some of the challenges in this dataset caused by various noise sources. The effect of common noise is shown in Fig. 4.8a. Generally, these noises can be caused by common references such as muscle movements, eye blinking, etc. Further, some files in this dataset contain several electrodes with low or constant values that do not have any information referring to dropout data, as shown in Fig. 4.8b, represented in time and frequency domains, where channel 17 is a dropout channel. In the frequency domain, the dropout channels can be easily detected based on their power level. These channels have a larger or lower power level compared to other channels. Since these channels do not contain useful information, they should be detected and removed from the data. Figure 4.8c shows unusual spikes that may be caused by electrode disconnection, causing a square-like shape with a large amplitude. Figure 4.8d shows two channels with high spikes and unusual shapes that are similar to dropout data. Finally, the dropout segment is represented in Fig. 4.8e.

4.2.2 The Preprocessing Technique Based on Patient-Specific Reference

The steps of this method are shown in Fig. 4.7. At first four electrodes, electrodes of interest (EOI) are chosen using the following criteria:

1. The selected electrodes are not among the dropout electrodes.
2. It is preferred to choose the electrodes that are close to the lobe area in the brain where the seizure takes place. Research shows that the behavior of the electrodes in the seizure onset zone (SOZ) varies in the time and frequency domains [35, 138–143].
3. The selected electrodes should cover a large area.
4. The selected electrodes should include at least two of the seizure onset zone electrodes.

In the re-referencing step, two approaches to examine the impact of the reference electrode on the results. The first one is the average of four EOI referring to CAR, where earlier studies frequently used this methodology. In the second method, the reference is selected based on the EOI and is thus called the patient-specific re-reference method. The patient-specific reference method selects the reference electrode based on the following:

1. The reference should be selected from the same side of the EOI. For removing common noise that is dependent on the side (left or right) of the scalp such as ocular artifacts, the reference needs to be selected from the same side of the EOI.
2. The reference electrode must be chosen far away from the EOI in order to prevent losing primary data during subtraction.

Figure 4.6 shows some examples of EOI and the selected reference electrode. A total of 142 electrodes, comprising 21 scalp electrodes, 10 depth electrodes, 44 strips, 64 grid, 1 ECG, 1 EMG, and an EOG, are employed to record the data for this patient. Therefore, HR1, HR10, TBc4, and FL3 are chosen as EOI and are indicated in blue; GH1 is the reference electrode and is represented in red.

After electrode selection and segmentation, as shown in Fig. 4.7, the following steps are performed to clean each segment. A notch filter with a frequency of 50 Hz is used to remove the power line noise. Then, if the system detects dropout data and they are more than 20% of the data, the segment will be removed, and the next segment needs to be processed. Otherwise, the process will move on to the next step. If the system identifies a high spike with a peak-to-peak value greater than a threshold, the system moves on to the next step, where several conditions must be investigated. If the high spike occurs in all channels, based on what is discussed in the previous section, these spikes can be due to common noise. To process that, the

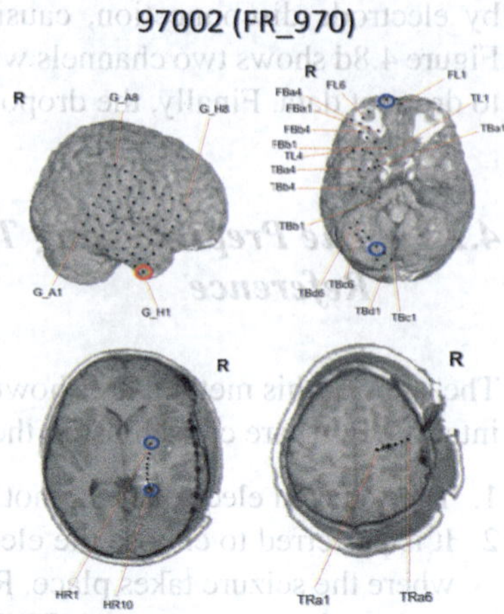

Fig. 4.6 Examples of EOI and the reference electrode

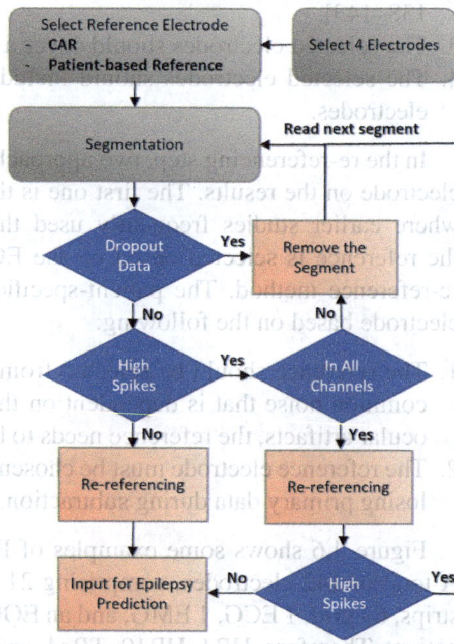

Fig. 4.7 The preprocessing method based on re-referencing [135]

system will go to the next step. If a high spike is seen in one, two, or three electrodes, in this case, the high spike occurs due to some other reasons such as disconnection of electrodes and scalp. These types of noises should be removed from the data. Then, the segment consisting of these spikes will be removed. Finally, common noise removal based on the two mentioned methods is used to remove the common noise and scalp reference. Note that the threshold for detecting the high spike is highly dependent on the data. For this dataset, based on monitoring the data, the threshold is selected to be 5000 μV. The threshold depends on the dataset.

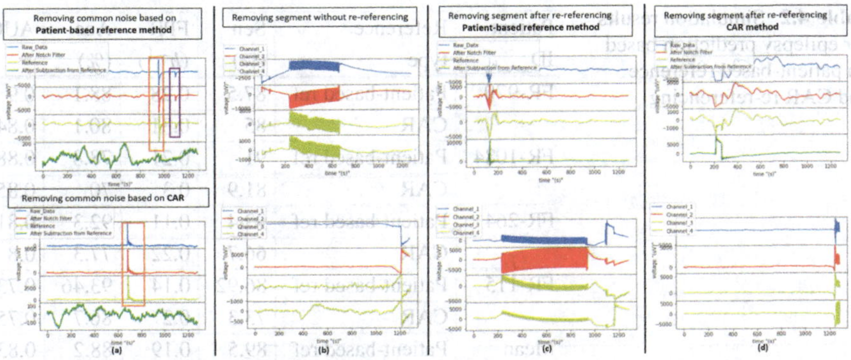

Fig. 4.8 Some examples of preprocessing steps, (**a**) Removing common noise, (**b**) Removing the segment consisting of spikes in one, two, or three channels, (**c**) Removing the spikes after the patient-based re-referencing method, (**d**) Removing the spikes after the CAR method

4.2.3 Performance Evaluation and Results

Figure 4.8 demonstrates the effectiveness of the preprocessing technique based on patient-based reference as it proceeds in different preprocessing steps. In Fig. 4.8a and b, the patient-based reference method and CAR method are represented in the first and second columns, respectively. Figure 4.8a is an example of removing high spikes using re-referencing based on patient-based reference and CAR methods. The high spike is detected in all channels which can be due to common noise. In Fig. 4.8a, raw data, signal after notch filter, the reference, and signal after re-referencing are represented for one channel. As can be seen, the high spike is removed after re-referencing. In Fig. 4.8b, however, the high spikes are detected in all channels, as they are not due to common noise, and these spikes still are not removed even after re-referencing based on the two mentioned methods. Therefore, the segment should be removed. In Fig. 4.8c and d, a high spike is detected in one, two, or three channels. In this case, the spikes are not due to common noise and mainly are due to electrode disconnection. Thus, the segment needs to be removed without any extra processing.

To investigate the effect of reference electrodes on the iEEG data, an LSTM model is used for epilepsy prediction, and the accuracy is applied as the main metric for comparison. The model is based on that in [26] but differs in the used activation functions and the number of electrodes. The model consists of data augmentation, feature extraction, classification, and postprocessing. In the data augmentation, the input signal is split into 5-second nonoverlapping segments. Subsequently, two mentioned techniques are used to clean the data. In the feature extraction step, 52 features in the time and frequency domains are extracted from each segment. In the classification, a two-layer LSTM, dropout, dense, and output layers are used. The ReLU and sigmoid activations are used in dense layers, the binary cross-entropy function is selected as the lost function, and the adaptive moment estimation (Adam)

Table 4.2 Simulation results for epilepsy prediction based on patient-based reference and CAR re-referencing

Patient ID	Reference type	Sen (%)	FPR (h^{-1})	Acc (%)	AUC
FR-970	Patient-based ref	87.5	0.28	88.1	0.9
	CAR	85	0.31	80.1	0.84
FR-1084	Patient-based ref	91	0.21	78.8	0.88
	CAR	81.9	0.3	70	0.85
FR-264	Patient-based ref	92.4	0.11	92.3	0.81
	CAR	66.7	0.22	77.3	0.8
FR-115	Patient-based ref	86.92	0.14	93.46	0.73
	CAR	73.3	0.2	86.7	0.75
Mean	Patient-based ref	89.5	0.19	88.2	0.83
	CAR	76.7	0.26	78.5	0.81

is used as the optimizer. The leave-one-out cross-validation (LOOCV) technique is used to evaluate the performance of the system.

Table 4.2 compares the performance of various alternative implementations of the mentioned preprocessing techniques based on patient-specific reference and CAR methods. The average results indicate that the method based on patient-based reference obtained a higher sensitivity of 89.5, an AUC Score of 0.83, and an accuracy of 88.2 as well as a lower FPR of 0.19 compared to the method based on CAR. This confirms that the selected reference based on the patient-specific method can be the better option compared to the average reference. This improvement can be as a result of maintaining the majority of the data by choosing the appropriate reference and eliminating common noise, while the bulk of the data in CAR will be deleted in addition to the common noise since the reference is identical to that of other channels.

4.3 Conclusion

This chapter has evaluated various denoising techniques for removing the artifacts from iEEG and sEEG signals. Simulation results showed that the WT with universal threshold using hard thresholding performs the best in removing motion artifacts and denoising of sEEGs. For iEEG denoising and removing ocular artifacts, WT with SURE Shrink using hard thresholding is preferred. We have also obtained the details for two preprocessing techniques that can remove various artifacts from iEEG data, including a novel patient-specific re-referencing method to remove common noise. Simulation results indicate that the epilepsy prediction accuracy using an LSTM model based on patient-specific method has a better performance with an average AUC of 0.83 compared to the same model when using the common average reference (CAR) method with an average AUC of 0.81.

Chapter 5
Efficient Epileptic Seizure Prediction Model Using European Dataset with Seizure Type Evaluation

5.1 LSTM

By retaining the information for a longer length of time, long short-term memory networks (LSTM), a specific type of recurrent neural networks (RNN), are able to learn long-term dependencies in the given data sequences, avoiding the vanishing gradient problem of RNN [144]. For the first time, Hochreiter and Schmidhuber [145] introduced LSTM networks. These networks are frequently employed to solve a variety of classification issues including time-series data, voice, audio, text, and biological signals, etc. A fundamental LSTM cell, Fig. 5.1, which has three gates for regulating the information flow from one cell state to another, serves as the basis for the LSTM architecture.

The forget gate, input gate, and output gate are some of these gates. In a given data sample, the forget gate determines whether a particular piece of information should be maintained or forgotten. In the cell state C_t, it takes into account the current input signal x_t and the previous output sequence y_t to produce an output f_t value between 0 and 1, where 0 denotes entirely forgetting the knowledge and 1 denotes completely remembering it. Similar to this, by combining the output of one tanh activation layer with the output of another, the output gate determines the flow of the proportion of information y_t in C_t at the output of the LSTM cell. The following equations mathematically explain how three gates of an LSTM cell operate to produce output y_t in cell state C_t: [50r3]

$$f_t = \sigma(W_f.[y_{t-1}, x_t] + b_f) \tag{5.1}$$

$$i_t = \sigma(W_i.[y_{t-1}, x_t] + b_i) \tag{5.2}$$

$$C_t = \sigma(W_C.[y_{t-1}, x_t] + b_C) \tag{5.3}$$

S. M. Varnosfaderani et al., *Hardware Software Co-design of Epileptic Seizure Prediction Systems*, Analog Circuits and Signal Processing,
https://doi.org/10.1007/978-3-031-95357-6_5

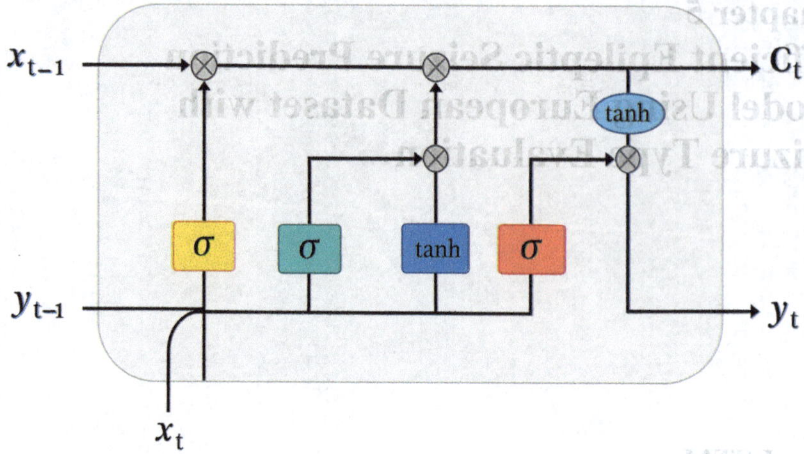

Fig. 5.1 LSTM cell

$$f_t = \tanh(W_C.[y_{t-1}, x_t] + b_C) \tag{5.4}$$

$$C_t = \sigma(f_t * [C_{t-1}, i_t] * C_t) \tag{5.5}$$

$$o_t = \sigma(W_o.[y_{t-1}, x_t] + b_o) \tag{5.6}$$

$$y_t = \sigma(o_t * \tanh C_t). \tag{5.7}$$

5.2 European iEEG Dataset

The European iEEG Epilepsy dataset is among the largest available datasets for epileptic seizure applications [136]. The dataset provides comprehensive details of each patient, including age, gender, disease history, surgical process and outcomes, the onset zone of seizures, and electrode type and placement. For each patient, the EEG onset and clinical onset are reported. EEG onset is generally assigned several seconds to several minutes earlier than the clinical onset (which is related to the first clinical signs). Each interictal file consists of a one-hour recording with a two-hour gap before and after EEG onset time to avoid overlapping. In our study, the preictal samples are extracted one hour before seizure onset with a five-minute gap from the EEG onset to avoid overlapping. Twenty-six patients from this dataset are used to evaluate the performance of the system. More details about this dataset can be found in [135] and [137].

5.2.1 Challenges in the European iEEG Dataset

In our study, as presented in [135], we utilized the European dataset for the first time to assess the effectiveness of a common average reference technique to mitigate the noise introduced by the common scalp reference. We encountered several challenges, including identifying and removing dropout channels and segments, reducing the noise caused by the scalp reference used for data recording, managing large noise spikes that are possibly induced by electrode disconnection, and recording subclinical seizures. The dataset consists entirely of raw data without any preprocessing, posing difficulties in cleaning the data from various types of noise. Although the dataset includes iEEG data, which is less susceptible to noise, a scalp EEG electrode serves as a reference electrode. Therefore, the noise generated from the scalp electrode manifests in the iEEG data as common noise.

Figures 5.1, 5.2, 5.3, and 5.4 present some types of noise exist in the European dataset. Figure 5.2 shows an example demonstrating the influence of common noise on EEG channels in time domain, caused by the movement of the reference electrode, motion artifacts, such as muscle movements, and eye blinking. Figure 5.3 illustrates the difference between the dropout channel (channel 15) and other channels in time and frequency domain. Several files in this dataset have multiple electrodes with low or constant values, which do not represent EEG data and thus can be considered as dropout data. Based on their power level, the dropout channels may be quickly identified in the frequency domain. Figure 5.4 displays a large spike in one channel, represented by red color, which can be due to disconnection of the scalp electrode. Figure 5.5 displays a large spike in one channel causing a large amplitude on all of the channels, which can be due to common noise.

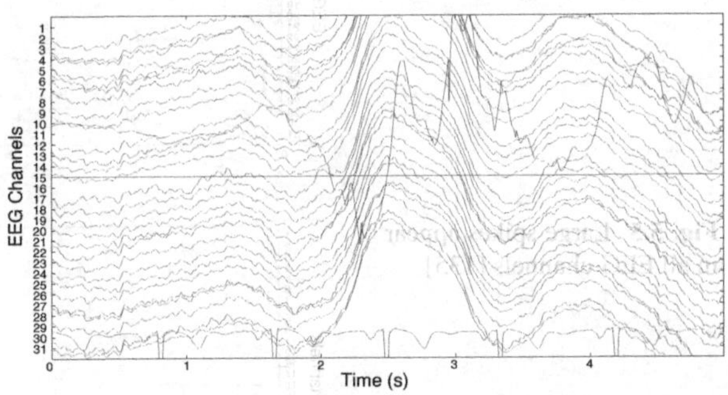

Fig. 5.2 An example of common noise in iEEG dataset [135]

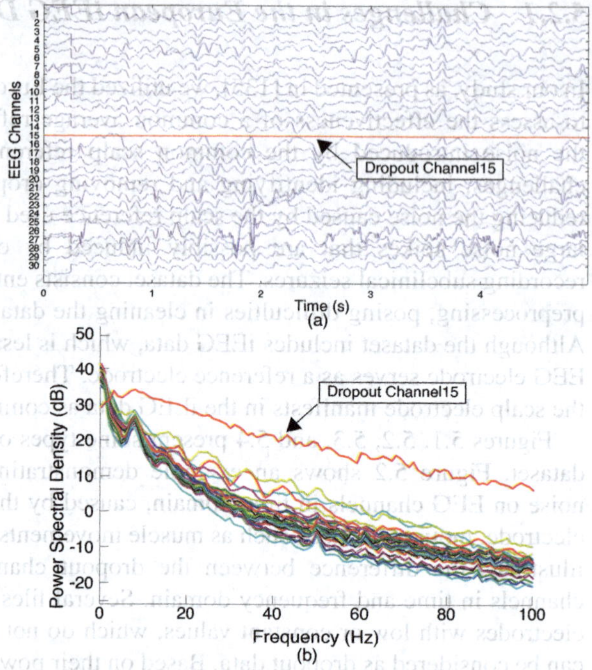

Fig. 5.3 The representation of dropout channels in time and frequency domain. (**a**) Time-domain representation where the red channel consists of dropout data. (**b**) Frequency-domain representation where the red channel consists of dropout channel [135]

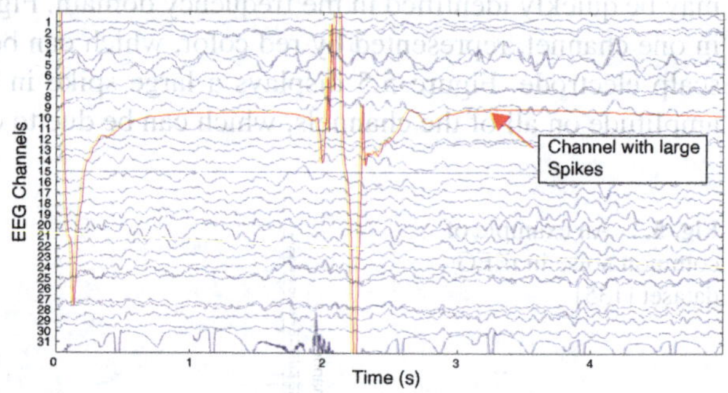

Fig. 5.4 An example of an electrode with a large spike, which is not due to common noise [135]

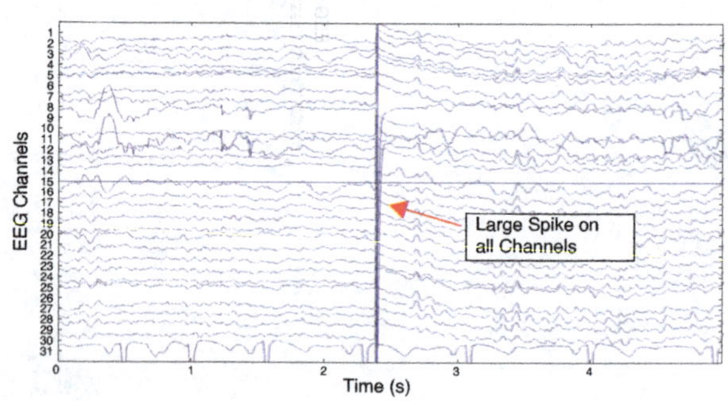

Fig. 5.5 Large spikes appear in all EEG channels [135]

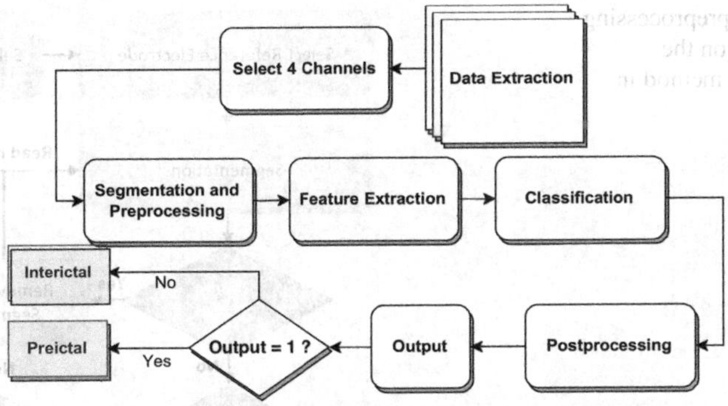

Fig. 5.6 The epilepsy prediction system for the European dataset

5.3 The Epilepsy Prediction Model Based on a Two-Layer LSTM

The epilepsy prediction system, designed for the European dataset, is illustrated in Fig. 5.6. As shown in the figure, the details of the model are described as follows:

1. **Data Extraction**: A one-hour EEG data with five-minute gap from the EEG onset is considered to extract preictal samples to avoid overlapping with the ictal state. Every interictal file lasts for one hour, where we consider the two-hour intervals before and after the EEG onset time to extract the interictal files.
2. **Channels Selection**: We select four EEG channels based on the following criteria: (1) proximity to the brain's lobe area where seizures start, (2) coverage of a large brain area, (3) inclusion of at least two seizure onset zone electrodes, and (4) exclusion of dropout electrodes. Furthermore, we assume that experts accurately assigned the origin electrode, as included in the dataset. More details can be found in our previous research [135].
3. **Segmentation and Preprocessing**: Each 60-minute raw iEEG clip is divided into 5-second nonoverlapping segments, and every six consecutive segments are grouped together as an LSTM input. Each preictal group is assigned a label of one, and each interictal group is assigned a label of zero. After that, the preprocessing is applied to clean each segment based on CAR method, as shown in Fig. 5.7 [135]. As depicted in the figure, once the four electrodes are selected, the segments are cleaned based on the following steps. (1) The power line noise is eliminated using a notch filter set to a 50 Hz frequency. (2) If the algorithm finds a dropout data (more than 20% of the segment), then the segment is discarded and the following segment is processed. (3) The system detects large spikes in the data when the peak-to-peak value surpasses a specific threshold. If a large spike is present in all channels, it is considered common noise, which is then removed using the re-referencing method described in [135]. In contrast, if a large spike is observed in one, two, or three electrodes, then these electrodes

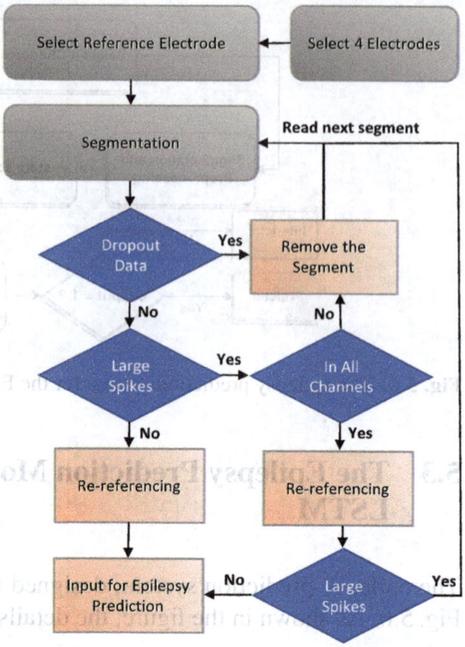

Fig. 5.7 The preprocessing method based on the re-referencing method in [135]

are likely disconnected and the segment is discarded. Note that the threshold for identifying large spikes is set to 5000 μV for this dataset.

4. **Feature Extraction**: Six features in time domain and seven features infrequency domain are extracted for each channel. The time-domain features consist of energy distribution, deviation, peak-to-peak values, and the number of zero crossings, as well as two statistical moments: *skewness* and *kurtosis*. The frequency-domain features include the spectral power intensity, which is extracted from the EEG data for 7 frequency bands from 0.1 Hz to 100 Hz using PyEEG library [146], including 0.1–4 Hz, 4–8 Hz, 8–12 Hz, 12–30 Hz, 30–50 Hz, 50–80 Hz, 80–100 Hz.
5. **Classification**: A two-layer LSTM is used to classify the data into preictal and interictal classes. Five different LSTM sizes are used to investigate the effect of the model's size on the results. Model 1 is the simplest model that consists of two layers with 16 memory units, while Model 5 consists of two layers with 256 memory units. The models are followed by a fully connected layer with an output of 30 to 100 units using the "ReLU" activation function and a final dense layer using "Sigmoid" activation function. The output is assigned to one for preictal samples and zero for interictal samples with a threshold of 0.5. More details about each model are shown in Fig. 5.8. For all models, the binary cross-entropy function is selected as the loss function, and the adaptive moment estimation (Adam) with a learning rate of 0.001 is used as the optimizer. The leave-one-out cross-validation (LOOCV) technique is utilized to evaluate the performance of the system. In the LOOCV technique, in each round, one preictal file and N

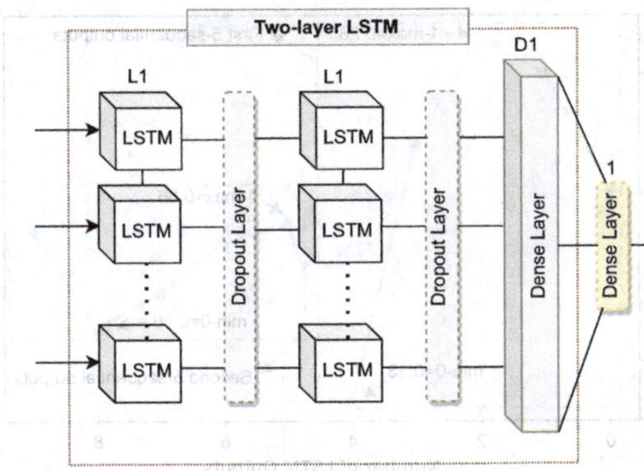

(a)

Details for each Model

Model	L1	D1	Num. of Parameters
Model 1	16	30	7,069
Model 2	32	50	20,901
Model 3	64	100	69,577
Model 4	128	100	237,257
Model 5	256	100	867,529

(b)

Fig. 5.8 (**a**) The structure of the classifier, which incorporates an LSTM model. (**b**) Details of each model used for evaluation

interictal files (to cover almost all data for the test part which is dependent on the total number of interictal files) are separated for testing, and the rest of the samples are used for training. For instance, we may have 50 interictal samples and 5 preictal samples for a single subject. Thus, we perform five test and training rounds for this patient, where we exclude one preictal sample and ten interictal samples in each round. The models are trained with an equal number of interictal and preictal samples from the training samples to address the unbalanced dataset issue. Dropout is applied to the first and second layers with a factor of 10% and 30%, respectively, to prevent overfitting. The number of epochs utilized for model training ranged from 100 to 500. In this case, the early stopping technique is implemented to stop the training before convergence to avoid the overfitting [147].

6. **Postprocessing**: Three postprocessing methods are applied, including *arithmetic mean, k-of-n* [75], and minimum distance (*MD*) [26] to assign an output to ten-sequence outputs from LSTM. In our study, we chose the values of k to be 5 and n to be 10. The MD method aims to determine the similarity between network outputs and either interictal or preictal samples. To make a decision about a sequence of outputs and assign a final output, the method assesses whether the output is closer to zero (representing interictal samples) or closer to one (representing preictal samples). In the following, more details about MD method are provided:
   - Identify Extremes: Determine the minimum and maximum of the n-sequence output of the LSTM network.
   - Calculate Distances: Calculate the distance between the minimum and zero (stored as dis-min) and the distance between the maximum and one (stored as dis-max). The objective of this step is to assess the similarity of the output to both interictal and preictal samples.

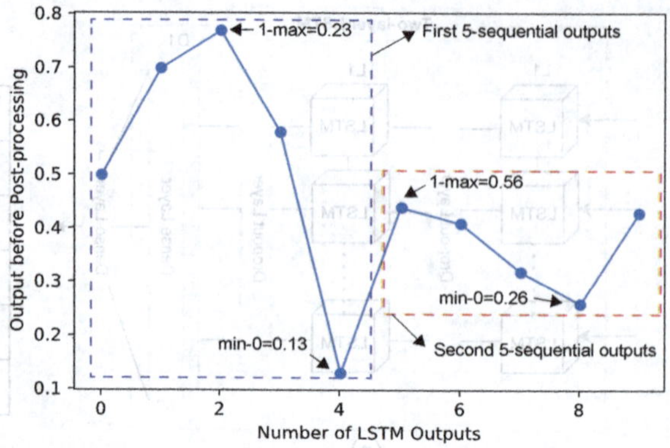

Fig. 5.9 An example of the MD method

– Comparison: Compare dis-min and dis-max to decide the final output. If dis-max is lower than dis-min, indicating similarity to preictal samples, assign the output as one. Otherwise, assign it as zero, representing interictal samples.

Figure 5.9 illustrates an example of how the MD technique operates, where each point on the y-axis represents an LSTM output, while the x-axis shows the sequence of outputs. In this system, the output is zero for interictal samples and one for preictal samples. Each set of five-sequence outputs of the LSTM ($n = 5$) is condensed into one output during postprocessing. In this example, the first and second sequence outputs are interictal samples. For both sequences, the dis-min values of 0.13 and 0.26 are lower than the dis-max values of 0.23 and 0.56, respectively. Therefore, the final output is correctly assigned to zero. However, for the first sequence output, the output is assigned to one using the mean and k-of-n method.

5.3.1 Evaluation of the Effect of Postprocessing on the Results

To assess the performance of this epileptic prediction model, simulation results for 26 patients from the iEEG European dataset are presented, utilizing metrics such as sensitivity, false prediction rate (FPR), accuracy, and area under the receiver operator characteristic (ROC) curve (AUC). The results for each patient and the average results for each model are presented in Tables 5.1, 5.2, 5.3, 5.4, and 5.5 for models 1–5, respectively. In these tables, bold values indicate the average values across the 26 patients for each metric. As shown in Table 5.1, for all patients except Pat-970, Pat-958, Pat-620, Pat-635, Pat-264, Pat-583, Pat-548, Pat-115, and Pat-442, the mean method has the best AUC, and k-of-n has the worst AUC. In these patients, the MD has the best AUC compared to the mean and k-of-n methods. The average results indicate that the mean method in the postprocessing presents the best AUC compared to MD and k-of-n. Pat-253 has the lowest AUC compared to other

Table 5.1 Evaluate the effect of postprocessing method on the results based on model 1

Model 1	Sen (%)			FPR (h^{-1})			Acc (%)			AUC		
Patient	Mean	MD	kofn	Mean	MD	kofn	Mean	MD	kofn	Mean	MD	kofn
Pat-970	81.52	82.7	78.7	0.23	0.22	0.20	77.8	78.4	80	0.87	0.87	0.79
Pat-1096	87.6	85.8	86.7	0.13	0.15	0.12	75.9	74.1	69.9	0.949	0.890	0.882
Pat-1084	86.1	77.3	84.2	0.19	0.19	0.16	82.0	80.3	83.5	0.94	0.89	0.84
Pat-958	88.3	91.7	83.3	0.12	0.16	0.10	87.8	84.8	89	0.926	0.928	0.864
Pat-922	92.5	90.8	90.0	0.23	0.21	0.22	80.7	81.8	81.2	0.927	0.918	0.825
Pat-273	72.2	60.8	69.9	0.28	0.28	0.24	72.1	71.6	75	0.863	0.808	0.708
Pat-253	100	100	97.2	0.38	0.44	0.29	65.7	59.6	73.3	0.76	0.753	0.663
Pat-862	100	100	95.9	0.01	0.03	0.01	99.2	97.2	98.8	1	0.975	0.975
Pat-620	100	100	100	0.29	0.25	0.27	74.7	72.7	76.3	0.98	0.995	0.87
Pat-1146	86.1	86.1	86.1	0.12	0.12	0.12	87.9	87.9	88.2	0.903	0.89	0.84
Pat-375	100	100	100	0.065	0.1	0.04	94.5	91.8	96.7	0.98	0.92	0.98
Pat-818	90.0	89.1	82.7	0.39	0.4	0.38	63.5	62.7	64.3	0.784	0.76	0.727
Pat-1073	91.5	89.8	79.5	0.32	0.27	0.28	70.5	74.5	72.7	0.846	0.842	0.758
Pat-916	80.6	75	80.6	0.12	0.22	0.11	87.3	77.8	88.1	0.91	0.793	0.847
Pat-635	72.2	75	66.7	0.24	0.3	0.19	76.1	70.7	79.2	0.767	0.793	0.703
Pat-565	94.4	94.4	94.4	0.21	0.2	0.19	81	82.1	82.6	0.95	0.94	0.86
Pat-264	100	100	95.85	0.32	0.36	0.32	70.7	68	71.1	0.84	0.895	0.82
Pat-1077	75.4	79.8	73.3	0.24	0.22	0.22	76.4	78.2	77.89	0.875	0.853	0.758
Pat-384	93.0	94.5	87.9	0.3	0.30	0.29	74.1	74.2	73.6	0.972	0.968	0.81
Pat-139	73.2	78.8	71.4	0.32	0.35	0.29	68.5	66.3	70.9	0.814	0.766	0.708
Pat-583	91.7	93.3	85	0.2	0.2	0.2	85.8	86.7	82.4	0.948	0.966	0.826
Pat-1125	86.1	91.7	86.1	0.35	0.34	0.34	67.4	68	70.1	0.813	0.798	0.758
Pat-548	66.7	66.7	65.7	0.37	0.34	0.35	64.9	66.4	65.1	0.764	0.792	0.652
Pat-590	71.7	70.5	70.6	0.02	0.02	0.01	84.6	83.9	84	0.981	0.966	0.841
Pat-115	75	72.9	75	0.18	0.11	0.19	80.3	85.7	79.9	0.813	0.818	0.783
Pat-442	72.8	70.5	70.6	0.16	0.16	0.12	83	82.3	85	0.828	0.841	0.785
Mean	**85.7**	**85.3**	**83**	**0.22**	**0.23**	**0.20**	**78.18**	**77.2**	**79.2**	**0.885**	**0.870**	**0.803**

patients, where the data are from a 37-year-old woman whose epilepsy began when she was 18 years old.

Several hypotheses could explain the suboptimal performance of the model. First, the data are recorded over 10 days, during which the patient underwent more than 70 subclinical seizures in a condensed time frame. Although the subclinical seizures are identified and eliminated from the data, it is posited that unreported subclinical seizures may persist, potentially influencing both the training and testing phases. Second, when the surgery was performed on this patient, the procedure resulted in Engle class II [148], which may have been caused by the incorrect identification of the original electrodes used in the simulations. A low number of samples to train the system can be another reason for poor performance for this patient. Pat-635, Pat-548, and Pat-818 also have an AUC lower than 0.8. The data for Pat-635

Table 5.2 Evaluate the effect of postprocessing method on the results based on model 2

Model 1	Sen (%)			FPR (h^{-1})			Acc (%)			AUC		
Patient	Mean	MD	kofn	Mean	MD	kofn	Mean	MD	kofn	Mean	MD	kofn
Pat-970	80.0	80.5	79.5	0.31	0.29	0.28	70.6	72.1	73.2	0.85	0.84	0.76
Pat-1096	89.5	88.7	89.5	0.14	0.14	0.13	86.0	86.3	87	0.954	0.952	0.882
Pat-1084	79.1	77.3	77.4	0.2	0.21	0.18	80.3	79.2	81.8	0.94	0.916	0.768
Pat-958	86.7	85	85	0.16	0.17	0.14	84.5	83.4	85.6	0.89	0.89	0.852
Pat-922	92.5	89.2	90.8	0.25	0.24	0.23	79.9	79.6	80.5	0.916	0.874	0.838
Pat-273	60.8	65.3	60.8	0.35	0.32	0.31	65	67.4	68.6	0.818	0.818	0.65
Pat-253	94.5	94.4	94.5	0.32	0.32	0.3	70.2	70.45	72.5	0.9	0.847	0.823
Pat-862	87.5	87.5	87.5	0.02	0.02	0.01	97.6	97.6	98	0.995	0.98	0.935
Pat-620	100	100	100	0.2	0.23	0.19	82	79.6	83.6	0.975	0.985	0.91
Pat-1146	80.0	83.1	74	0.15	0.15	0.15	84.4	84.9	84.4	0.837	0.827	0.797
Pat-375	100	95.5	100	0.11	0.12	0.1	90.7	89	91.2	0.93	0.905	0.95
Pat-818	84.2	87.8	77.9	0.33	0.35	0.31	68.7	67.4	670	0.766	0.757	0.736
Pat-1073	79.38	81.1	70.9	0.26	0.27	0.20	75	73.7	79	0.836	0.818	0.75
Pat-916	75	86.1	75	0.09	0.16	0.06	90.1	84	91.9	0.91	0.863	0.833
Pat-635	64.1	55.8	61.3	0.36	0.30	0.33	66.2	70.5	68.3	0.84	0.723	0.75
Pat-565	94.4	91.7	94.4	0.14	0.12	.13	87.1	88.2	87.9	0.94	0.94	0.9
Pat-264	100	100	100	0.4	0.4	0.39	64.5	64.1	64.9	0.845	0.935	0.81
Pat-1077	80.1	84.3	77.9	0.2	0.23	0.22	76	77.5	78.1	0.908	0.905	0.78
Pat-384	94.5	94.5	92.9	0.28	0.27	0.27	75.7	76.6	76.3	0.97	0.954	0.828
Pat-139	64.1	71.4	64.1	0.32	0.35	0.30	68.0	65.6	69.5	0.728	0.686	0.696
Pat-583	91.7	96.7	80	0.25	0.22	0.25	83.2	87.4	77.3	0.944	0.928	0.84
Pat-1125	90.3	91.7	90.3	0.33	0.33	0.32	69	69.5	70	0.778	0.78	0.788
Pat-548	63.6	62.8	58.8	0.37	0.35	0.34	63.5	64.0	62.6	0.784	0.77	0.623
Pat-590	64.1	73.9	66.4	0	0.02	0	82	85.7	82	0.977	0.957	0.816
Pat-115	73	68.8	73	0.11	0.09	0.09	86.3	86.6	87.8	0.81	0.725	0.823
Pat-442	69.7	72.9	67.4	0.23	0.2	0.19	76.6	79.7	79.5	0.821	0.826	0.767
Mean	**82.3**	**83.3**	**80.4**	**0.23**	**0.23**	**0.21**	**77.8**	**78.1**	**78.9**	**0.88**	**0.862**	**0.804**

are collected from a 63-year-old woman whose epilepsy began when she was 30 years old. For this patient, the majority of reported seizures are related to the unclassified seizure category which may be the reason for the poor performance. More than 70% of reported seizures for Pat-548 and Pat-818 fall into the categories of simple partial and unclassified seizure types, which might negatively affect the system's effectiveness. Model 1 performs well for Pat-862 with the highest AUC of 1, the lowest FPR of 0.01, and the highest accuracy of 99.2%. The data are collected from a 16-year-old female experiencing seizures for four years. For this patient, no subclinical seizure is reported which can be one reason for having a good performance. Note that for this patient, 100% of seizures are the secondarily generalized seizures.

Table 5.3 Evaluate the effect of postprocessing method on the results based on model 3

Model 1	Sen (%)			FPR (h^{-1})			Acc (%)			AUC		
Patient	Mean	MD	kofn	Mean	MD	kofn	Mean	MD	kofn	Mean	MD	kofn
Pat-970	77.7	76.6	76.6	0.25	0.23	0.22	75.0	76.6	77.6	0.86	0.84	0.77
Pat-1096	87.5	88.8	85.7	0.15	0.15	0.14	85.4	85.3	86.5	0.951	0.94	0.856
Pat-1084	80.9	70.0	79.2	0.2	0.2	0.19	80.1	79.4	80.6	0.896	0.91	0.8
Pat-958	83.3	90.0	80	0.13	0.16	0.12	87.0	84.2	86.9	0.912	0.892	0.84
Pat-922	92.5	89.2	90	0.23	0.22	0.21	80.9	81.1	82.2	0.912	0.881	0.845
Pat-273	60.6	60.4	51.7	0.24	0.26	0.23	75.4	68.5	70.8	0.838	0.783	0.65
Pat-253	100	97.2	100	0.45	0.49	0.38	59.0	55.5	65.4	0.9	0.803	0.807
Pat-862	95.9	95.9	95.9	0.02	0.04	0.02	98.4	96.4	98.4	0.99	0.97	0.97
Pat-620	100	100	100	0.22	0.25	0.2	80.4	78	82.4	0.99	0.995	0.9
Pat-1146	65.6	65.4	62.6	0.08	0.1	0.06	90	88	91.0	0.86	0.85	0.783
Pat-375	100	100	100	0.11	0.15	0.11	90.1	90.5	94.4	0.91	0.905	0.945
Pat-818	85.5	89.2	74.6	0.37	0.40	0.34	65.6	62.5	66.7	0.753	0.758	0.7
Pat-1073	88.0	87.9	81.2	0.28	0.28	0.25	73.4	73.4	75.7	0.842	0.816	0.782
Pat-916	91.7	83.3	91.7	0.24	0.26	0.21	77.7	75.4	79.7	0.88	0.817	0.85
Pat-635	66.7	66.7	63.9	0.3	0.28	0.28	70	71.5	71.8	0.8	0.753	0.68
Pat-565	94.4	88.9	88.9	0.14	0.12	0.13	86.5	88.2	87.0	0.94	0.93	0.88
Pat-264	91.7	91.7	91.7	0.4	0.44	0.4	63	59.4	63.3	0.775	0.82	0.76
Pat-1077	86.7	82.2	76	0.28	0.25	0.27	73.3	75.6	73.1	0.813	0.833	0.743
Pat-384	94.5	94.5	89.5	0.27	0.27	0.26	76.9	76.9	76.6	0.964	0.96	0.816
Pat-139	66.7	75.2	59.4	0.31	0.37	0.26	69.3	64.2	73	0.722	0.716	0.67
Pat-583	81.7	88.3	78.3	0.24	0.24	0.22	79.0	82.3	78.2	0.956	0.916	0.782
Pat-1125	88.2	91	88.2	0.36	0.36	0.35	66.7	66.6	67.3	0.815	0.763	0.765
Pat-548	60.9	59.1	58.8	0.37	0.33	0.32	62.0	63.3	63.5	0.737	0.728	0.633
Pat-590	64	66.3	59	0.02	0.05	0.02	80.1	80.7	78.2	0.953	0.943	0.784
Pat-115	73	70.9	70.8	0.08	0.06	0.08	88.3	89.1	87.4	0.805	0.83	0.813
Pat-442	64.4	62.5	63.4	0.16	0.18	0.14	82.1	80.6	83.9	0.821	0.83	0.75
Mean	**82.4**	**82**	**79.1**	**0.23**	**0.24**	**0.21**	**77.5**	**76.7**	**78.5**	**0.869**	**0.853**	**0.791**

As shown in Table 5.2, the mean method performs the best for the majority of patients, compared to other postprocessing methods. The worst performance is reported by Pat-139 with an AUC of 0.728. For this patient, 60% of the seizures are in the simple partial category which might be related to the poor performance. As the AUC decreases from 0.814 in Model 1 to 0.728 in Model 2, the low number of samples for training the system could be another reason. The reported AUC for Pat-590 is also among the highest reported AUC. This data is for an 18-year-old male who experienced a seizure when he was 11 years old. Several reasons might cause good performance. The patient's surgical outcome is classified as Engle class I, confirming that those electrodes used for simulation are correctly selected from the seizure onset zone. Further, all seizures are among the secondarily generalized and complex partial seizures. As displayed in Table 5.3, the mean method in the

Table 5.4 Evaluate the effect of postprocessing method on the results based on model 4

Model 1	Sen (%)			FPR (h^{-1})			Acc (%)			AUC		
Patient	Mean	MD	kofn	Mean	MD	kofn	Mean	MD	kofn	Mean	MD	kofn
Pat-970	82.8	80.6	79.4	0.27	0.29	0.25	73.6	72.5	76.1	0.88	0.86	0.78
Pat-1096	91.3	89.7	89.5	0.17	0.17	0.16	83.6	84.08	84.6	0.949	0.942	0.869
Pat-1084	80.8	78.9	80.9	0.18	0.19	0.16	81.6	80.7	83.6	0.926	0.93	0.822
Pat-958	95.0	86.7	86.7	0.16	0.15	0.15	85.1	84.9	85.5	0.91	0.936	0.858
Pat-922	92.5	90.8	89.2	0.23	0.22	0.22	81.3	81.3	81.1	0.911	0.889	0.837
Pat-273	68.6	75.4	66.5	0.21	0.2	0.17	78	78.3	81.6	0.815	0.808	0.748
Pat-253	97.2	88.9	88.9	0.27	0.34	0.22	75.0	68.4	79.1	0.88	0.767	0.833
Pat-862	95.9	100	95.9	0.02	0.05	0.01	98.4	96	98.8	1	0.97	0.975
Pat-620	100	100	100	0.17	0.15	0.15	84.9	86.1	86.5	0.985	0.995	0.925
Pat-1146	83.3	80.3	77.5	0.11	0.09	0.09	89	90	89.7	0.873	0.84	0.843
Pat-375	100	100	100	0.1	0.15	0.1	91.6	87.3	91.2	0.95	0.94	0.93
Pat-818	86.3	87.6	85.01	0.36	0.37	0.34	66.5	65.7	68.1	0.793	0.784	0.757
Pat-1073	88.02	84.38	82.86	0.31	0.26	0.28	71.1	75.1	73.3	0.83	0.834	0.776
Pat-916	86.1	83.3	80.6	0.14	0.19	0.12	85.8	81.2	87.8	0.907	0.833	0.847
Pat-635	58.3	50	55.6	0.21	0.20	0.21	77.4	44.9	76.9	0.737	0.713	0.673
Pat-565	88.9	86.1	86.1	0.16	0.13	0.16	84	86.8	84.3	0.93	0.88	0.85
Pat-264	79.2	83.4	70.9	0.23	0.25	0.23	77.4	75.8	77	0.835	0.89	0.745
Pat-1077	93.6	87.1	89.2	0.3	0.3	0.26	72.6	71.3	75.5	0.878	0.795	0.8175
Pat-384	92.9	92.9	92.9	0.24	0.26	0.22	78.9	77.5	80.4	0.97	0.956	0.85
Pat-139	71.4	71.4	71.4	0.31	0.35	0.23	69.1	65.6	71.8	0.734	0.684	0.716
Pat-583	91.7	95	86.7	0.25	0.22	0.24	83.2	86.6	81.6	0.962	0.954	0.816
Pat-1125	86.1	82.7	86.1	0.32	0.32	0.31	69.5	69.2	70.9	0.79	0.75	0.778
Pat-548	57.3	59.5	56.3	0.34	0.35	0.33	61.8	62.4	61.9	0.758	0.727	0.616
Pat-590	62.8	71.4	60.3	0.05	0.09	0.05	78.9	81.3	77.6	0.95	0.961	0.776
Pat-115	75	75	75	0.21	0.18	0.16	78.2	81.1	82.4	0.83	0.823	0.798
Pat-442	70.7	68.7	69.6	0.19	0.16	0.16	80.4	82.5	82.8	0.791	0.809	0.77
Mean	**83.7**	**82.7**	**80.9**	**0.21**	**0.22**	**0.19**	**79.1**	**77.6**	**80.4**	**0.876**	**0.857**	**0.808**

postprocessing achieves the best performance compared to other methods. The results for Pat-139 are still among the worst results compared to other patients. The reported results for Pat-862 are still among the best-reported results among all patients. Further, as shown in Table 5.4, the mean method still achieves the best performance.

As indicated in Table 5.5, the best results are observed for Pat-862, with the highest accuracy and sensitivity both reaching 100%, the maximum AUC of 1, and the lowest FPR of zero. As mentioned before, there are no reported subclinical seizures for this patient, potentially contributing to the excellent performance. This underscores the influence of seizure type and data characteristics on the system's efficacy.

Table 5.5 Evaluate the effect of postprocessing method on the results based on model 5

Model 1	Sen (%)			FPR (h^{-1})			Acc (%)			AUC		
Patient	Mean	MD	kofn	Mean	MD	kofn	Mean	MD	kofn	Mean	MD	kofn
Pat-970	84.3	81.6	80.5	0.26	0.26	0.23	73.8	74.8	77.0	0.85	0.86	0.79
Pat-1096	84.6	88.5	81.8	0.15	0.15	0.14	84.7	85.7	85.9	0.934	0.948	0.841
Pat-1084	82.7	80.9	80.9	0.22	0.19	0.20	78.8	81.2	79.9	0.906	0.93	0.802
Pat-958	86.7	86.7	81.7	0.19	0.20	0.20	81.4	80.6	80.8	0.884	0.878	0.812
Pat-922	95	88.3	92.5	0.33	0.34	0.32	74	72.1	74.4	0.903	0.852	0.802
Pat-273	60.2	68.7	51.4	0.22	0.21	0.2	76.3	78.2	77.6	0.82	0.82	0.66
Pat-253	75	58.3	63.9	0.21	0.21	0.18	78.9	78.5	79.9	0.853	0.787	0.727
Pat-862	100	100	100	0	0	0	100	100	100	1	1	1
Pat-620	100	100	100	0.16	0.16	0.15	85.3	85.7	86.9	0.975	1	0.93
Pat-1146	77	77.0	77	0.12	0.13	0.12	86.9	86.6	87.4	0.873	0.857	0.827
Pat-375	100	100	100	0.06	0.17	0.05	95.1	85.7	95.6	0.965	0.865	0.975
Pat-818	86.4	86.2	83.9	0.35	0.36	0.33	67.4	69.5	68.5	0.811	0.793	0.754
Pat-1073	82.9	77.9	77.7	0.27	0.26	0.23	73.7	73.7	77	0.846	0.82	0.774
Pat-916	80.6	83.3	77.8	0.18	0.18	0.15	82.2	82.47	83.9	0.9	0.837	0.813
Pat-635	63.9	61.1	58.3	0.47	0.37	0.38	59.07	62.6	61.4	0.68	0.66	0.603
Pat-565	91.7	86.1	91.7	0.15	0.14	0.13	85.7	86.3	87.6	0.92	0.88	0.89
Pat-264	83.4	87.5	79.2	0.3	0.38	0.3	71.5	64.8	71.1	0.79	0.79	0.735
Pat-1077	80.5	78.4	78.4	0.16	0.18	0.13	84	81.8	85	0.91	0.883	0.825
Pat-384	94.5	92.9	94.5	0.24	0.25	0.24	79.5	59.1	79.5	0.966	0.972	0.852
Pat-139	75.3	75.3	75.3	0.35	0.42	0.33	65.6	59.6	68.2	0.76	0.754	0.714
Pat-583	100	100	96.7	0.2	0.2	0.2	90	90	88.34	0.966	0.93	0.9
Pat-1125	90.3	87.5	88.2	0.31	0.31	0.3	71.1	71.5	72.1	0.783	0.808	0.793
Pat-548	51.3	51.4	50.4	0.289	0.27	0.25	61.2	62.3	62.8	0.7167	0.68	0.627
Pat-590	63.8	67.5	57.5	0.02	0.04	0.013	80.67	82	78.2	0.947	0.929	0.781
Pat-115	60.4	68.8	60.4	0.08	0.13	0.07	85.3	83.7	86.2	0.708	0.8	0.77
Pat-442	64.4	60.3	63.4	0.15	0.15	0.13	83.1	82.4	86.6	0.776	0.778	0.751
Mean	**81.3**	**80.5**	**78.6**	**0.21**	**0.22**	**0.19**	**79.0**	**77.7**	**80**	**0.863**	**0.850**	**0.798**

The impact of postprocessing techniques, namely mean, MD, and K-of-N, on evaluation metrics (sensitivity, FPR, accuracy, and AUC) is summarized in Fig. 5.10. Notably, the AUC values demonstrate that the mean method outperforms MD and K-of-N techniques across all models during postprocessing. Consequently, the mean technique is adopted during the postprocessing stage.

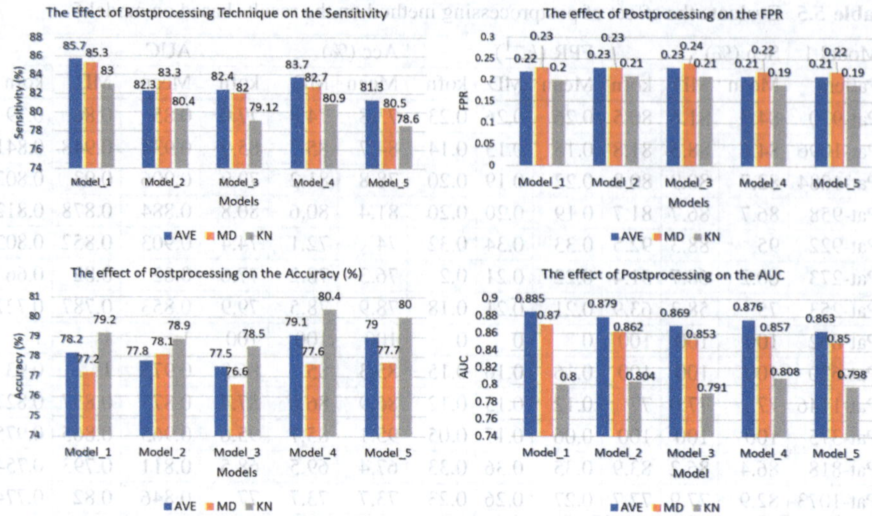

Fig. 5.10 A summary of the effect of various postprocessing techniques on the system's performance

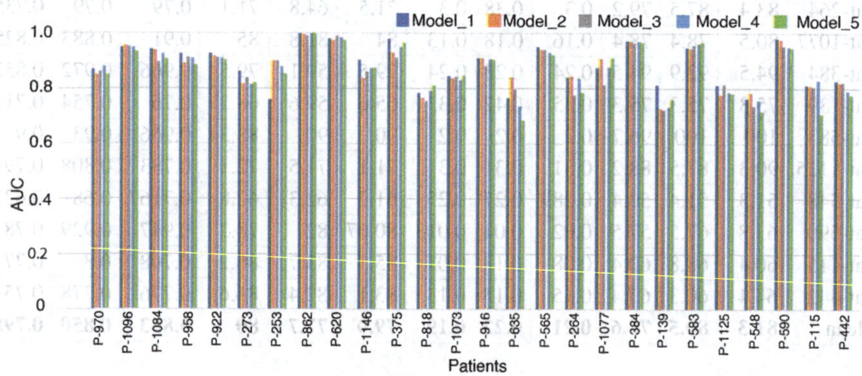

Fig. 5.11 The effect of the model's complexity on the system's performance for each patient

5.3.2 *Evaluation of the Effect of Model Complexity on the Results*

As outlined in the preceding section, the mean method in postprocessing demonstrated superior performance compared to MD and K-of-N methods. Consequently, to assess the impact of model complexity on system performance, the average outcomes for the mean method are compared across different models. The comparative results for each patient are shown in Fig. 5.11. Notably, an increase in model size enhances results for certain patients, attributed to the complexity of their data,

Fig. 5.12 The impact of model's complexity on the AUC

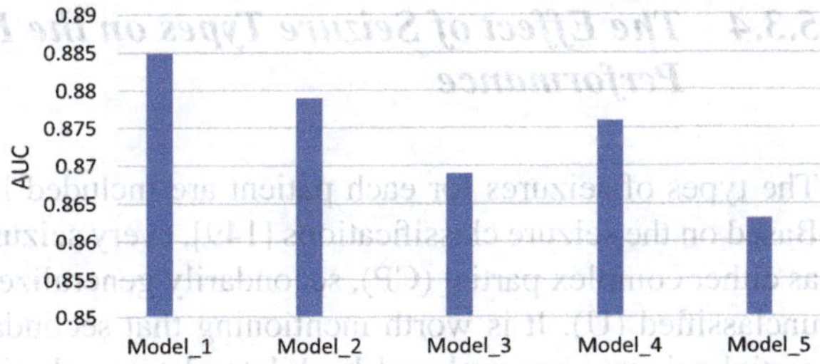

thereby indicating that a larger model yields improved outcomes. However, for some patients, enlarging the model diminishes performance, possibly due to overfitting, resulting in a consistent average performance as illustrated in Fig. 5.12. Further, the figure indicates that augmenting the LSTM size does not markedly influence average results. Although a more complex model might be expected to yield better outcomes, simulation results revealed minimal changes, likely due to averaging results across 26 patients. While individual patient results may exhibit noticeable changes with increased model size, the overall average remains relatively constant. The comparative analysis further reveals that model 1, employing the mean method in postprocessing, achieved the highest AUC compared to other model sizes.

5.3.3 Comparative Analysis with Prior Models

Table 5.7 presents a comparative analysis of epilepsy prediction models. It is crucial to highlight that, unlike studies conducted on the Melbourne dataset, MIT dataset, American Society, and Bonn datasets that typically reported results for the number of patients between 3 and up to 23, our results in this table are specifically for 26 epileptic seizure patients from the European dataset. The assessment of our model is based on four key metrics, namely sensitivity, FPR, accuracy, and AUC. In contrast to some systems that report only one or two metrics, our comprehensive evaluation provides a more robust understanding of the system's performance. Note that this work marks the first instance of applying the iEEG European dataset, and this model demonstrates commendable performance with low complexity. Importantly, we introduce an effective and simple model that utilizes only four electrodes, making it suitable for both implanted and wearable technology applications. This resulted in a classifier with fewer parameters—just 7,069—and a lower level of complexity. However, many reported models tend to be more complex, with a high number of trainable parameters. As an example, we note that the lists of trainable parameters for models in [24, 32–34] are, respectively, 18,354, 20,125, 2,200,000, and 69,129,612, which clarifies the significant variation in complexity across the range of suggested approaches. Therefore, this model strikes a balance between complexity and accuracy and thus has the potential to be implemented in resource-constrained devices, including wearables and implants.

5.3.4 *The Effect of Seizure Types on the Prediction Performance*

The types of seizures for each patient are included in the European iEEG dataset. Based on the seizure classifications [149], every seizure in this dataset is categorized as either complex partial (CP), secondarily generalized (SG), simple partial (SP), or unclassified (U). It is worth mentioning that secondarily generalized seizures and partial seizures are replaced by bilateral tonic-clonic seizures and focal seizures, respectively, by the ILAE from 1981 classification in 2017 [149]. In this section, we investigate the impact of the seizure type on the prediction results. In Table 5.6, the AUC and sensitivity for model 1 using the mean method in the postprocessing stage are presented for each patient. The last four columns on the right side indicate the percentage of each seizure type present in the data for each patient in the simulation.

Table 5.6 System's performance and the percentage of seizure types used for each patient in the iEEG European Dataset

Model 1 Patient	Sen (%)	AUC (%)	SG (%)	CP (%)	SP (%)	U (%)
Pat-970	81.52	0.87	0	93.3	0	6.7
Pat-1096	87.6	0.949	0	88.9	0	11.1
Pat-1084	86.1	0.94	0	0	0	100
Pat-958	88.3	0.926	20	60	0	20
Pat-922	92.5	0.927	0	0	20	80
Pat-273	72.2	0.863	0	75	0	25
Pat-253	100	0.76	33.3	33.3	33.3	0
Pat-862	100	1	100	0	0	0
Pat-620	100	0.98	0	50	50	0
Pat-1146	86.1	0.903	0	66.7	0	33.3
Pat-375	100	0.98	0	0	50	50
Pat-818	90.0	0.784	28.6	0	57.1	14.3
Pat-1073	91.5	0.846	0	0	100	0
Pat-916	80.6	0.91	0	0	0	100
Pat-635	72.2	0.767	0	33.3	0	66.7
Pat-565	94.4	0.95	0	0	100	0
Pat-264	100	0.84	0	50	0	50
Pat-1077	75.4	0.875	0	25	50	25
Pat-384	93.0	0.972	0	20	0	80
Pat-139	73.2	0.814	0	40	60	0
Pat-583	91.7	0.948	0	0	80	20
Pat-1125	86.1	0.813	0	0	50	50
Pat-548	66.7	0.764	11.1	0	88.9	0
Pat-590	71.7	0.981	28.6	71.4	0	0
Pat-115	75	0.813	0	0	50	50
Pat-442	72.8	0.828	0	100	0	0

For example, Pat-970 does not experience SG or U seizures, while 93.3% and 6.70% are CP and SP seizures, respectively (Table 5.7).

In the following results, sensitivity is used as the primary metric to evaluate the system's effectiveness, as seizure type can significantly impact the sensitivity measure. The average sensitivity for each patient based on seizure type is represented in Fig. 5.14. As indicated in the figure, the seizure type affects the sensitivity, where the lowest reported sensitivity is related to the unclassified seizure type for Pat-273, while the complex partial type has a sensitivity of 87.87%. Furthermore, the reported sensitivity of unclassified and simple partial categories for each patient is often lower than the reported sensitivity of complex partial and secondary generalized seizure types. Moreover, the lowest sensitivities of SG, CP, SP, and U are 66.65%, 72.83%, 50%, and 25%, respectively. This confirms that the epileptic prediction system can have better performance on those patients with CP and SG seizures compared to those patients suffering from SP and U seizures. The average sensitivity for each seizure category is shown in Fig. 5.13. The average results indicate that CP seizures have the highest sensitivity of 89.9%, while the lowest sensitivity is related to SP seizures. Thus, the seizure types can explain the reasons behind the system's good or poor performance for each patient (Fig. 5.14).

5.4 Discussion and Future Work

In this chapter, we have presented a model that stands out for its good performance and low complexity. We achieved this by selectively using four electrodes to simplify the model architecture. This resulted in a classifier with fewer parameters—just 7,069—and a lower level of complexity. Twenty-six patients were used in our simulation framework. However, including more patients can lead to more substantial and impactful results. Future work aims to expand upon these findings by incorporating a larger patient sample size. Note that our findings highlight how important it is to have accurate information about the origin electrodes, seizure onset zones, and electrode placement of each patient during the electrode selection step. Consequently, databases utilized for this purpose must comprehensively obtain these pertinent details. As a result, the use of some publicly accessible datasets like the Melbourne Dataset and the American Epilepsy Society Dataset may be limited since they do not contain these essential details.

This model operates on a patient-specific basis, requiring tailored retraining and testing procedures for each individual. Looking ahead, we aspire to refine our model to achieve patient-independent, streamlining processes for broader applicability and efficiency. Generally, there is a trade-off between complexity and performance. Maintaining a balance between these factors will be essential as we work to improve our model's efficacy and complexity. Since we aimed to present a model with lower complexity, our model's performance has to be enhanced by decreasing FPR. Using CNNs for feature extraction instead of handcrafted features is one way to make improvements. However, it is crucial to acknowledge that such advancements may

Table 5.7 Reported results on epilepsy prediction algorithms

Ref.	Dataset	Preprocessing	Feature extraction	Classification	Postprocessing	Sensitivity %	FPR %	Accuracy %	AUC
[65]	Melbourne	Downsampling	STFT	CNN	–	87.85	–	–	
[66]	Melbourne CHB-MIT	Notch filter	STFT	Convolutional epileptic Network (CESP)	–	78.11 88.21	0.14 0.27	–	
[67]	Melbourne CHB-MIT	Continuous wavelet Transform (CWT)	–	Semi-dilated convolution Convolution network (SDCN)	–	88.45 98.9	– –	– –	0.883 0.928
[39]	Melbourne	–	EEG scalogram	Multichannel	–	91.15	–	–	0.924
				Vision transform (MViT)					
[85]	Melbourne	–	–	CNN	–	85.2–86.3	–	–	0.914–0.933
[86]	Melbourne	–	–	MLP	–	–	–	–	0.815
[56]	Melbourne	–	Time and frequency	Random forest, adaptive	–	–	–	–	0.844
[87]	Melbourne	–	Signal energy	Decision tree, kNN	–	33.67	–	–	–
[88]	Melbourne	–	Signal energy,	Logistic regression	–	52.67	–	–	–
			Circadian profile						
[89]	Melbourne	–	EEG spectrogram, Circadian profile	CNN	–	77.36	–	–	–
[26]	Melbourne	Filtering	Statistical features in time domain, power in different frequency bands	Two-layer LSTM	K-of-N, MD, Mean	86.8	0.147	85.1	92
[59]	American Epilepsy Society	–	STFT	CNN	K-of-N method	75	0.21	–	

[60]	CHB-MIT	–	Time/frequency domain, wavelet transform, cross correlation, graph theory	Two-layer LSTM	–	99.28–99.84	0.02–0.11	–	–
[58]	CHB-MIT	–	Deep convolutional autoencoder	BiLSTM	–	99.72	0.004	99.66	–
[73]	CHB-MIT	–	STFT + CNN	LSTM	K-of-N method	98.21	0.13	–	–
[63]	CHB-MIT	Filtering	Combination of common spatial statistics	CNN	Kalman filtering	92.2	0.12	90	–
[62]	CHB-MIT	–	STFT	CRGNN	–	89	–	75.6	
[90]	CHB-MIT	–	Spectral power	SVM		98.7	0.04	–	–
[91]	CHB-MIT	–	Phase locking value	SVM		82.4	–	–	–
[92]	CHB-MIT	–	Mel frequency Spectral coefficient	Siamese NN		92.5	–	91.5	–
[93]	CHB-MIT	–	–	CNN		92	0.136	–	–
[94]	CHB-MIT	–	Spectral- Temporal feature	Graph convolution Network		95.5	0.109	–	–
[95]	CHB-MIT	–	EEG Spectrogram	LSTM		93	–	–	–
[96]	CHB-MIT	–	EEG Spectrogram	Residual network		89.3	–	92.1	–

(continued)

Table 5.7 (continued)

Ref.	Dataset	Preprocessing	Feature extraction	Classification	Postprocessing	Sensitivity %	FPR %	Accuracy %	AUC
[97]	CHB-MIT	–	Mel frequency	GNN		94.5	–	95.4	–
			Spectral coefficients						
[98]	CHB-MIT	–	–	Diated CNN		93.3	0.007	–	–
[99]	CHB-MIT	–	EEG	ViT		59.2–97	–	–	–
			Spectrogram						
[74]	Bonn	–	–	Stacked BiLSTM	–	89.21	0.06	91	
[100]	European	CAR-based method	Time and frequency	Two-layer LSTM	Mean	85.7	0.22	78.1	0.885

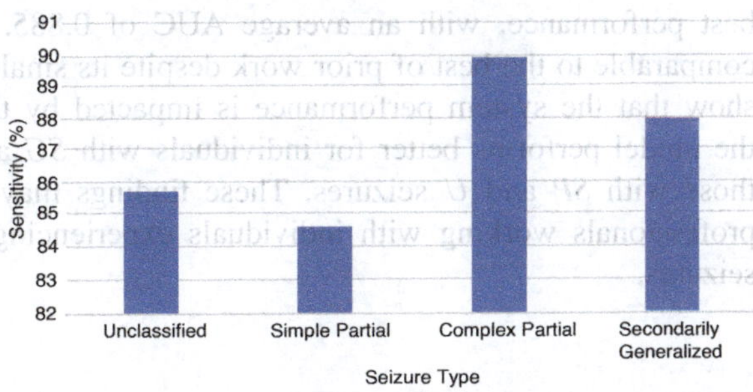

Fig. 5.13 The effect of various seizure types on the sensitivity

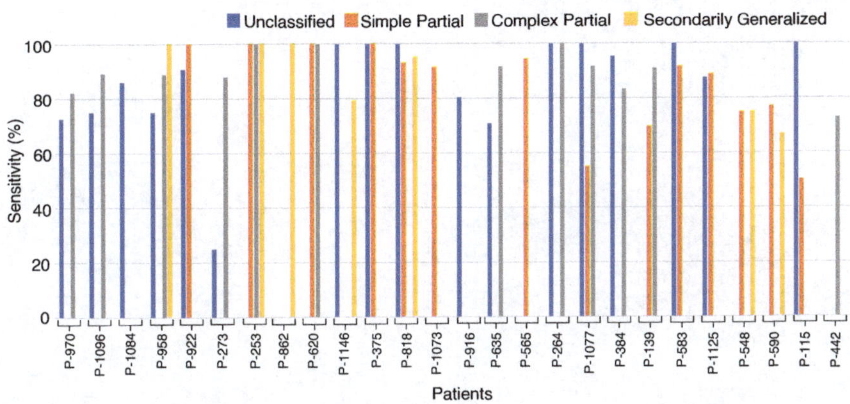

Fig. 5.14 The effect of seizure type on model's sensitivity for each patient

come at the expense of increased complexity. In this research, we also looked at how different seizure types affected the results. However, it is important to recognize that seizure type is just one aspect of a more complex picture. Incorporating a wider range of patient characteristics and illness background variables, such as age, gender, and seizure onset zone, among others, is crucial for thorough and detailed comparisons. We anticipate that these parameters will be incorporated into our studies in the future, which will enhance the scope and depth of our results.

5.5 Conclusion

We obtained the details of an epileptic seizure prediction system based on a two-layer LSTM and have conducted a detailed study, which is the first to utilize the large iEEG European dataset and analyze the impact of seizure type. We analyzed the results of five different model sizes and three different postprocessing methods on a large set of patients. Simulation results indicate that the model with its simplest structure in conjunction with the mean method for postprocessing achieves the best performance, with an average AUC of 0.885. The system performance is comparable to the best of prior work despite its small model size. The results also show that the system performance is impacted by the seizure type. Specifically, the model performs better for individuals with *SG* and *CP* seizures compared to those with *SP* and *U* seizures. These findings may offer a new perspective for professionals working with individuals experiencing different types of epileptic seizures.

Chapter 6
A Self-Aware Power Management Model for Epileptic Seizure Systems Based on Patient-Specific Daily Seizure Pattern

6.1 Hardware-Based Solutions

Current commercial solutions monitor EEG in real time to detect the seizure onset. For example, the *responsive neurostimulation system* (RNS) is an implantable device that detects the occurrence of certain types of seizures and provides stimulation to reduce the effect of a seizure [150]. Another device, called *Embrace Watch*, detects possible convulsive seizures by measuring sympathetic nervous system activity (without using EEG signals), but it fails to detect all convulsive seizures. There is a need not only to detect seizures but also to predict them before they happen so that the patients can take precautions, stop certain activities (such as driving, swimming alone, or climbing ladders), and/or reduce their effects [151].

Deep learning (DL) methods have become the cornerstone for modern artificial intelligence (AI) applications, especially for healthcare, due to the unprecedented achieved accuracy [168]. DL has been applied in a wide spectrum of healthcare applications, including predicting epileptic seizures [169], predicting heart attacks [69], finding tumors in MRI images [170], and forecasting COVID-19 cases [171]. Although DL methods can be designed to predict epilepsy with high accuracy [58], the high computational resources and memory bandwidth requirements are the key challenges in implementing DL models in low-power resource-constrained devices.

As illustrated in Fig. 6.1, an automated EEG-based, battery-powered system for epilepsy prediction includes various hardware blocks: electrodes, data acquisition and digitization, hardware accelerator, wireless communication, and power management. Such a system is desired to achieve continuous monitoring and accurate prediction while maintaining low cost and a long battery lifetime [172]. The number of channels is determined by the number of used electrodes. These channels are filtered and digitized by the data acquisition and digitization block and then processed by the hardware accelerator, which is a custom digital block that implements the DL model. The wireless communication block shares the

S. M. Varnosfaderani et al., *Hardware Software Co-design of Epileptic Seizure Prediction Systems*, Analog Circuits and Signal Processing,
https://doi.org/10.1007/978-3-031-95357-6_6

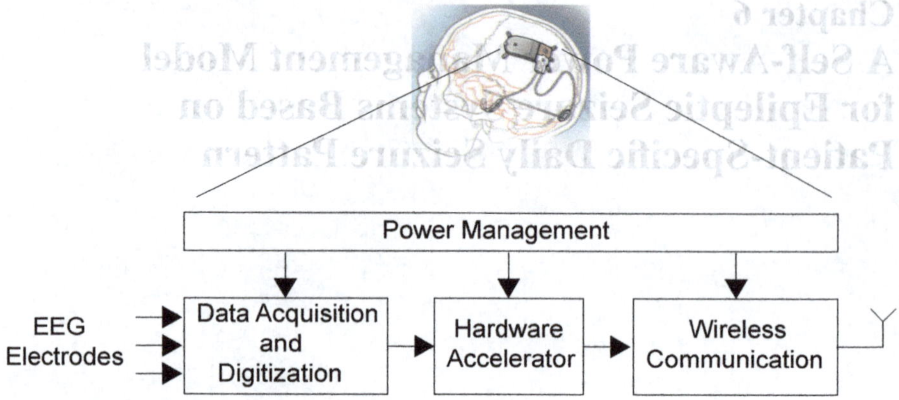

Fig. 6.1 A battery-powered system for predicting epileptic seizures

results with the patient as well as the medical professionals, who can specify the patient-specific patterns and adjust anti-seizure drugs accordingly. Finally, the power management module delivers power to all the blocks, ensuring low power consumption and a long battery lifetime.

Digital accelerators can be custom-made for deep neural networks (DNNs) to provide higher throughput, shorter latency, lower energy, and higher area efficiency [173]. Although digital accelerators provide better performance compared to GPUs, digital systems (including both GPUs and digital accelerators) are fundamentally limited in handling big data efficiently due to the separation of logic and memory (referred to as the Von Neumann bottleneck). Consequently, the system bandwidth is limited by the speed of accessing the data in the memory. Moreover, memory access requires at least 10x higher energy and longer delay compared to the actual computation in DNNs, specifically the multiply-accumulate (MAC) operation [174]. Recent advances in circuit design and memory technologies enable efficient computation through in-memory and near-memory computation, as well as more efficient computation in the analog domain and time domain. Further, it has been shown that computation can be done more efficiently in the analog domain [175] and time domain [176]. These technologies open the door for more innovation in hardware accelerators to achieve low-power and low-cost epilepsy prediction systems.

The use of a wearable accelerometer sensor in seizure detection systems is prevalent because it can detect sudden body movements, such as jerking motions, stumbling movements, and quick body falls, which are common phenomena for most seizure types, such as generalized tonic-clonic seizure (GTCS), tonic-clonic, myoclonic, and hypermotor. Accelerometers utilized in seizure detection systems are precise, three dimensional, and energy efficient. The majority of accelerometer applications in seizure detection emerge as wearable devices, such as the SmartWatch Inspyre, E4 wristband, Embrace wristband, NightWatch, and a variety of other gadgets created by well-known technology corporations. An accelerometer can be worn in a variety of places on the body, including the wrist, chest, back, and ankle [177]. Channel selection is another approach, which selects only the channels that contain the most important information, thereby lowering the complexity, computation, and power consumption by decreasing the amount of data being processed. In [58], a selection algorithm is implemented to select 10 out of 23 channels on average based on the highest variance entropy product. The results showed that applying such an algorithm decreased 34% of parameters and 50% of training time while maintaining the same accuracy [58]. In [178], an SVM model is tested using 2 to 6 out of 22 channels. The results showed that applying channel selection decreased more than 93% of the computational costs. Channel selection can be an effective technique for reducing computational complexity, but the algorithm itself can be complex.

The power consumption of the hardware accelerator can be reduced by network pruning to remove unimportant connections [179, 180], reducing SRAM voltages based on the level of fault tolerance caused by bit masking, and optimizing the data types [180]. Feature extraction, bit-wise design, and estimation of the activation functions are some other solutions to simplify the design and reduce the design cost.

In this chapter, we discussed various systems and hardware accelerators for epilepsy prediction and detection and compared them in terms of performance, size, and power consumption. Besides, according to the patient-specific daily seizure pattern, we obtain the details of a model that can switch to low-power mode when the chance of seizure incidence is minimal or zero. This work can help researchers to understand the design challenges for designing an energy-efficient, low-cost, and portable system for epileptic seizure prediction. More details are obtained in the following.

6.1.1 Prior Work on Hardware Implementations of Epilepsy Monitoring Systems

Table 6.1 summarizes the hardware implementations of various epilepsy monitoring systems. As depicted in the table, two commercial solutions are available for epileptic patients; RNS performs detection and requires surgery, while Embrace is a noninvasive commercial device. RNS, an FDA-approved therapeutic option, is one of the interventional treatment options for people with refractory and focal epilepsy. The RNS system is the first commercially closed-loop responsive brain stimulation device [181]. RNS is implanted inside the brain to monitor EEG signals, detect seizures, and respond by applying electrical pulses to the patient's brain to reduce the effect of the seizure [150, 182]. Although RNS reduces the seizure frequency and severity, it has a high rate of false detections, raising the question of how much of this effect is attributable to closed-loop suppression of seizure-related ictal activity [183]. In addition, RNS is bulky, has a limited number of channels,

Table 6.1 Comparison table between various hardware implementations of epilepsy monitoring systems

Ref.	Goal	Implementation	Data (Num of patient)	Sensitivity (%)	FPR (h^{-1})	ML model	Power consumption (Watt) and CLK (MHz)
[152]	Prediction	FPGA	sEEG (4) Private data	100	3.9	One-class SVM	– and 50(MHz)
[153]	Prediction	FPGA	sEEG Bonn	90	–	BSNN	–
[154]	Prediction	FPGA	iEEG UPen	90	–	FC-DNN	1.9 and 25(MHz)
[155]	Prediction	FPGA	sEEG (10) European epilepsy	77.3	0.04	RusBoosted	–
[156]	Prediction	FPGA	sEEG (8) CHB-MIT	–	–	–	–
[157]	Prediction	FPGA	sEEG (5) CHB-MIT	77.4	–	DL	0.013
[158]	Detection	FPGA	–	–	–	Generic algorithm	–
[159]	Detection	FPGA	sEEG (24) CHB-MIT	98.4	0.356	SVM	0.38 and 100 (MHz)
[160]	Detection	–	sEEG (24) CHB-MIT	95.1	3.8	Dual LSVM	2.73 uJ/class
[161]	Detection	TSMC 180 nm	iEEG (14) CHB-MIT	95.7	0.02	Dual LSVM	–

[162]	Detection	TSMC 180 nm	iEEG (4) Long-Evans rats	–	–	Linear least square (LLS)	0.028 and 3.125 (MHz)
[163]	Detection	FPGA TSMC 180 nm	Data of mice	–	–	FFNN	0.01589 and 8 (MHz)
[164]	Detection	TSMC 65 nm	iEEG (26)	83.7	–	Boosted tree	41.2 nJ/class
RNS	Detection	Commercial device	iEEG	–	–	–	–
Embrace– Embrace 2 [165–167]	Detection and Prediction	Commercial device Apple-Android -Smartphone	Non-EEG Noninvasive	–	High-FPR	–	–

and relies on basic hard thresholding with moderate seizure classification accuracy. Because of the power and size limits imposed by implanted devices, complex on-chip classification algorithms cannot be implemented [184].

Wearable sensors, such as the Embrace and Embrace 2 watches developed by MIT, are extremely valuable due to their precision in detecting epilepsy seizures. Embrace watch developed by Empathica is a wrist-worn wearable device that monitors various physiological signals, including heartbeat, temperature, and respiratory rate using EDA (electrodermal activity) sensor. The watch uses machine learning to detect patterns that may be associated with convulsive seizures [185, 186]. The reading is obtained from the watch and transmitted using a Bluetooth connection. The reading is then forwarded to a cloud server and database for storage. This is only possible if an internet connection is present. Users who have been registered to the appropriate user can view the readings of the main user who is wearing the watch via the cloud server and database. When an epileptic seizure occurs, the watch notifies the other users who are registered to the appropriate user. Embrace2 is the second generation with improvements in battery life, weight, and connectivity [166]. Ictal Care365, Neuroon, Ricola, Vigil-Aide, Epi-Care free, and Zephyr are some examples of the possible potential solution for seizure monitoring using biomarker detection systems, which are available in the market [167].

A real-time seizure prediction system is presented in [152] based on an EEG dataset, which has 11 seizures extracted from four different focal epilepsy patients and obtained by the University of Texas Health Center using surface implanted electrodes. The system uses a one-class support vector machine (SVM) and is implemented on Zynq-7000 XC7Z045 FPGA which has the requisite FPGA fabric and DSP slices to host the computationally demanding method. Data are converted into fixed point to improve utilization of the FPGA fabric and DSP slices. The FPGA implementation consumes about 8 MB of the total 20 MB of block RAM (BRAM). The results showed that the system can predict seizures 3.6 minutes prior to clinical onset with a 3.9 mean false prediction rate (FPR) per hour. A hardware implementation of bit-serial neural network (BSNN) for epileptic seizure prediction is reported in [153]. The system utilizes a bit-serial data processing unit along with a finite state machine (FSM) to process the one-bit data at a time at each clock cycle to reduce power consumption and reduce area while running at low speed. The system is implemented on an ALTERA Cyclone V FPGA using 3931 ALMs which constitutes about 7% of the Cyclone V A7 capacity. The system achieved an accuracy of 90% when tested on the Bonn dataset [187]. For neural network computing, parallel hardware architecture is typically employed to boost performance. However, in this design, low power and low cost are more important than high performance, so a bit-serial architecture-based data processing unit (DPU) is presented for neural network computing to reduce power and cost. An SRAM is used to store the neural network's weights. For bit-serial processing, the ALU employs a proprietary multiplier. In [154], a deep neural network (DNN), implemented in a Xilinx Zynq-7000 Zybo-7 FPGA, is used to predict seizures. The system is tested on the American Epilepsy Society Seizure Prediction Challenge dataset [188]. The results show that the system predicts seizures with a 74%

accuracy, 90% true positive rate, and power consumption of 1.9 W. In [155], the RusBoosted classifier is applied to classify the data, which is suitable for unbalanced data. One channel is used and chosen to reduce power usage and simplify the model. Four blocks from FPGA are used: BRAM, DSP48E, flip flop (FF), and lookup table with 19%, 52%, 20%, and 62% usage, respectively.

Due to its low-power-consumption and -area requirements, epileptic seizure research has generally focused on ASIC implementation. Chen et al. [162] obtained accuracy higher than 92% at 2.8 mW power with a total area of 13.47 mm^2. Similarly, [160] and [164] demonstrated on-chip seizure detection with sensitivities of 95.1% and 83.7% with total areas of 1 mm^2 at 2.73 J and 41.2 nJ per class. However, because these processors are not reprogrammable, ASIC-based wearable devices may face adaption and accuracy loss issues with a more diversified epileptic patient population.

Studies on FPGA-based seizure detection, on the other hand, have revealed partial reconfiguration, high-speed complicated feature extraction, and online training capabilities. Wang et al. [159] obtained 98.4% sensitivity on the Xilinx Zynq-7000 while utilizing 380mW of power at 100 MHz via FFT-based feature extraction. Similarly, [154] obtained 74% accuracy for seizure prediction on the Zynq-7000 with 1909 mW power. Tsou et al. [163] extracted 256-point FFT-based features on an FPGA and classified them on-chip at 1.589 mW with a total area of $1409 \times 1402\, um^2$ for the neural network chip. Huang et al. [189] demonstrated a hybrid approach in which online training is performed on the FPGA and seizure detection is performed on-chip. This system, however, has greater design costs, intermediate data transfer latency, and larger space needs. FPGA's high-performance capabilities come at the expense of increased power consumption and energy requirements. Therefore, designing a feasible and scalable, battery-powered, FPGA-based, seizure detection system is a major task.

6.2 Model Based on Patient-Specific Seizure Patterns

To reduce energy consumption, we propose a new generation of self-aware wearable or implantable systems based on the patient-specific daily seizure pattern. Epileptic seizures have been shown to have biases in distribution over time at various intervals that can be as long as 1 year or as short as 1 h [75]. In another word, the probability of seizure occurrence for each patient can follow a particular daily pattern. For each patient, there are some hours when the risk of having a seizure is high and certain hours when the risk is low or nearly zero. Therefore, we can find a daily seizure pattern for each patient. By analyzing the patient-specific daily seizure pattern, we can manage the power during the day for each patient. Therefore, during times of the day when seizures are not predicted or have a low chance of occurring, the system can switch to a low-power mode.

We determine and utilize the distribution of the number of seizures during the day, which could help to improve epilepsy systems to forecast patient-specific

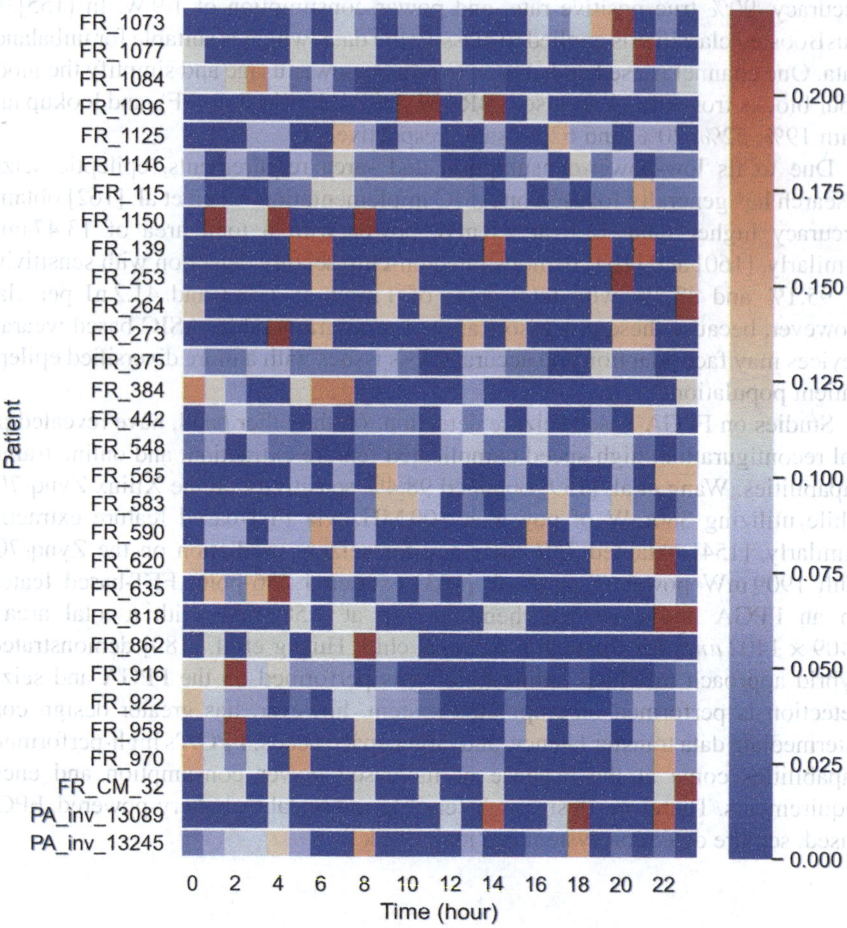

Fig. 6.2 Distribution seizures over the 24-hour period for the 30 patients

seizures through the addition of 24-hour-cycle information. The distribution of seizures over the 24-hour cycle for 30 patients from the European iEEG Epilepsy Dataset is shown in Fig. 6.2 [136, 137]. The color in each block represents the fraction of seizures that occurred during a certain hour. For example, patient FR_916 had 52 seizures, and thus a block with a value of 0.2 at 2 a.m. depicts that one-fifth of the seizures occurred around 2 a.m. In this graph, dark-blue blocks depict no-seizure intervals. Once the patient-specific daily seizure pattern is established, it can be incorporated into the prediction/detection system. The model for power management based on the patient-specific pattern is represented in Fig. 6.3. The patient's data are first read and digitalized. According to the daily seizure pattern for a patient, the power management unit reads the daily pattern and assigns the low-power or high-power mode; low-power mode is given to ratio values greater

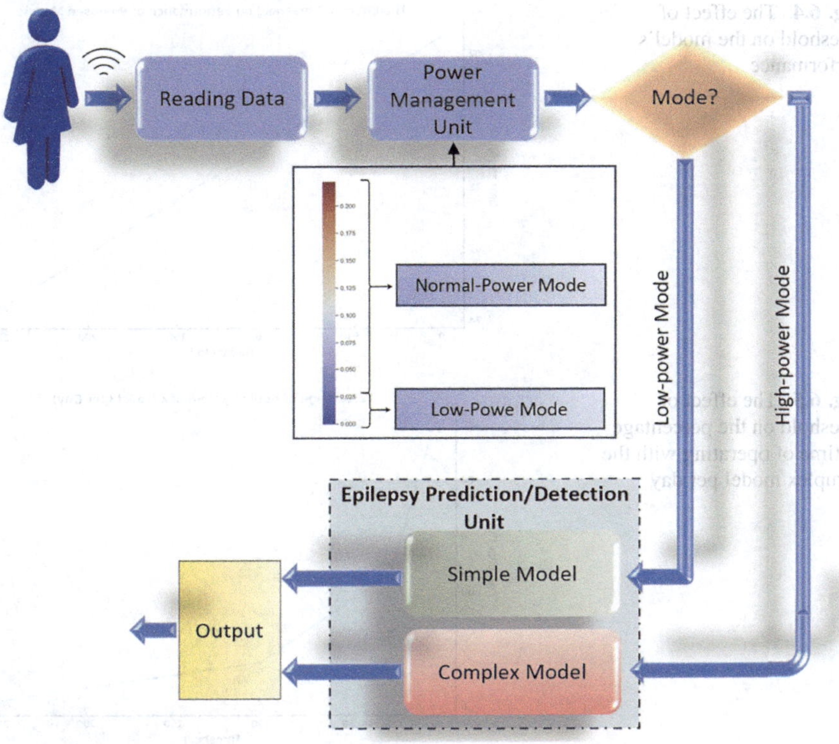

Fig. 6.3 Block diagram of the self-aware model

than 0.025, and high-power mode is given to ratio values greater than 0.025. The key to a patient's daytime power management will be this pattern. Afterward, a simple or complex ML model is employed for the low-power or high-power mode, depending on the assigned mode, respectively.

In this research, we employed the simple and complex detection models in [190, 191] where the classification can be done based on a basic set of features and a more intricate set of features, respectively. Based on the computed threshold in the train section, the two-level classifier in [190] determines whether the simple model is confident to classify the incoming data or the complicated one. Our suggested two-level self-aware classifier, on the other hand, changes from a basic model to a complicated model or vice versa depending on the patient-specific seizure pattern. Making the choice to move between two states in this situation does not need any additional calculations. The influence of threshold (T) on the percentage of time that the system operates in the complex model per day ($P_{Complex}$) and the influence of threshold (T) on the model's performance are shown in Figs. 6.4 and 6.5. As it can be seen, increasing the threshold from zero to 225 reduces the time of being in the complex model. As a result, the basic model, which has a lower Gmean and power consumption, has a greater impact on system performance. The energy (E)

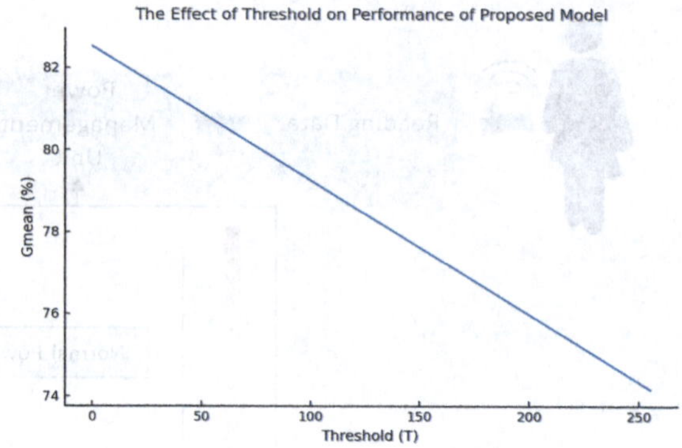

Fig. 6.4 The effect of threshold on the model's performance

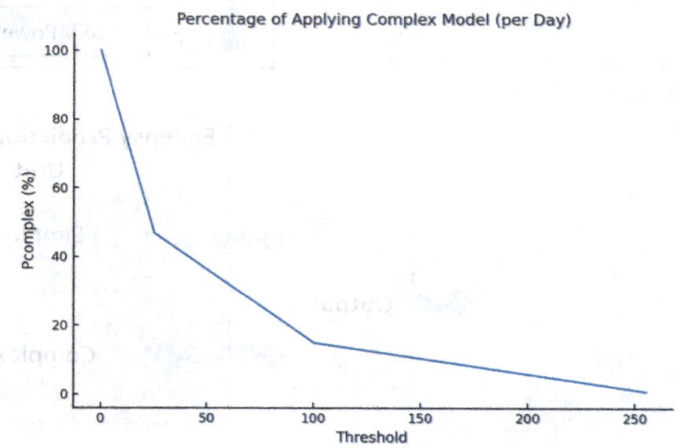

Fig. 6.5 The effect of threshold on the percentage of time of operating with the complex model per day

and Gmean for the power management model are calculated based on the following equations:

$$E = P_{Complex} \times (E_{Complex}) + P_{Simple} \times (E_{Simple}) \tag{6.1}$$

$$Gmean = \frac{1}{225}(T \times (Gmean_S) + (225 - T) \times (Gmean_C)), \tag{6.2}$$

where $P_{Complex}$ and P_{Simple} refer to the percentage of time per day using a complex and a simple model, respectively. $Gmean_C$ and $Gmean_{Simple}$ are the system's performance in the complex and simple models, respectively. T is the threshold which can be between 0 and 225.

The average results for 30 patients with $T = 25$ are obtained in Table 6.2. The energy consumption and Gmean for the self-aware model are calculated and compared with the simple and complex models. As it can be seen, this power management model could effectively reduce the power consumption by 49% compared to the complex model; however, the performance reduction is less than 1 %.

Table 6.2 Estimated results for epilepsy detection based on simple model, complex model, and power management model

Model	Low-power mode (Hours per day)	Gmean (%)	Energy (uJ)
Simple model	24	75.16	2.832
Complex model	0	82.53	31.464
This study	13	81.73	15.955

The results indicate that the power management model can greatly reduce power consumption for most patients while maintaining the performance.

6.2.1 Conclusion

This chapter performed a detailed analysis of the challenges and opportunities for designing low-power epilepsy prediction systems. Designing an efficient epilepsy prediction system requires various levels of software and hardware optimizations, including custom hardware accelerators, machine learning models, channel selection, and statistical analysis of patient-specific EEG data. We obtained the details of a self-aware power management model to reduce the power consumption for each patient based on the patient-specific pattern. This model is applicable for all types of epilepsy prediction and detection systems. The simulation results indicated that this model can effectively reduce the power consumption by 49 percent compared to the complex model without losing the performance. It is worth to mention that the switch between basic and complicated models is done based on the daily seizure pattern that is unique for each patient and does not require any additional computations. In the future, we need to include more parameters to get the decision for switching between low-power mode and high-power mode.

References

1. G.-F. Tian, H. Azmi, T. Takano, Q. Xu, W. Peng, J. Lin, N. Oberheim, N. Lou, X. Wang, H. R. Zielke, et al., An astrocytic basis of epilepsy. Nat. Med. **11**(9), 973–981 (2005)
2. T.A. Milligan, Epilepsy: a clinical overview. Am. J. Med. **134**(7), 840–847 (2021)
3. *Neurological Disorders: Public Health Challenges* (World Health Organization, Geneva, 2006)
4. R.S. Fisher, C. Acevedo, A. Arzimanoglou, A. Bogacz, J.H. Cross, C.E. Elger, J. Engel Jr, L. Forsgren, J.A. French, M. Glynn, et al., Ilae official report: a practical clinical definition of epilepsy. Epilepsia **55**(4), 475–482 (2014)
5. D. Thurman, E. Beghi, C. Begley, A. Berg, J. Buchhalter, D. Ding, D. Hesdorffer, W. Hauser, L. Kazis, R. Kobau, et al., ILAE commission on epidemiology. standards for epidemiologic studies and surveillance of epilepsy. Epilepsia **52**, 2–26 (2011)
6. D. Angeles, et al., Proposal for revised clinical and electroencephalographic classification of epileptic seizures. Epilepsia **22**(4), 489–501 (1981)
7. J.J. Falco-Walter, I.E. Scheffer, R.S. Fisher, The new definition and classification of seizures and epilepsy. Epilepsy Res. **139**, 73–79 (2018)
8. T. Tomson, T. Walczak, M. Sillanpaa, J.W.A.S. Sander, Sudden unexpected death in epilepsy: a review of incidence and risk factors. Epilepsia **46**(s11), 54–61 (2005)
9. E. Beghi, The epidemiology of epilepsy. Neuroepidemiology **54**(2), 185–191 (2020)
10. https://www.who.int/news-room/fact-sheets/detail/epilepsy/
11. W.A. Hauser, J.F. Annegers, L.T. Kurland, Incidence of epilepsy and unprovoked seizures in Rochester, Minnesota: 1935–1984. Epilepsia **34**(3), 453–458 (1993)
12. K.M. Fiest, K.M. Sauro, S. Wiebe, S.B. Patten, C.-S. Kwon, J. Dykeman, T. Pringsheim, D.L. Lorenzetti, N. Jetté, Prevalence and incidence of epilepsy: a systematic review and meta-analysis of international studies. Neurology **88**(3), 296–303 (2017)
13. E. Beghi, D. Hesdorffer, Prevalence of epilepsy—an unknown quantity. Epilepsia **55**(7), 963–967 (2014)
14. E.A. Kelvin, D.C. Hesdorffer, E. Bagiella, H. Andrews, T.A. Pedley, T.T. Shih, L. Leary, D.J. Thurman, W.A. Hauser, Prevalence of self-reported epilepsy in a multiracial and multiethnic community in New York City. Epilepsy Res. **77**(2–3), 141–150 (2007)
15. K. Kaiboriboon, P.M. Bakaki, S.D. Lhatoo, S. Koroukian, Incidence and prevalence of treated epilepsy among poor health and low-income Americans. Neurology **80**(21), 1942–1949 (2013)
16. N.E. Bharucha, E.P. Bharucha, A.E. Bharucha, A.V. Bhise, B.S. Schoenberg, Prevalence of epilepsy in the Parsi community of Bombay. Epilepsia **29**(2), 111–115 (1988)

S. M. Varnosfaderani et al., *Hardware Software Co-design of Epileptic Seizure Prediction Systems*, Analog Circuits and Signal Processing,
https://doi.org/10.1007/978-3-031-95357-6

17. P. Camfield, C. Camfield, Incidence, prevalence and aetiology of seizures and epilepsy in children. Epileptic Disord. **17**(2), 117–123 (2015)
18. V.L. Feigin, E. Nichols, T. Alam, M.S. Bannick, E. Beghi, N. Blake, W.J. Culpepper, E.R. Dorsey, A. Elbaz, R.G. Ellenbogen, et al., Global, regional, and national burden of neurological disorders, 1990–2016: a systematic analysis for the global burden of disease study 2016. Lancet Neurol. **18**(5), 459–480 (2019)
19. C. Cooper, Global, regional, and national burden of neurological disorders, 1990–2016: a systematic analysis for the global burden of disease study 2016. Lancet Neurol. **18**(4), 357–375 (2019)
20. https://www.healthpages.org/health-a-z/brain-injury-brain-functions/
21. Z. Altaf, M.A. Unar, S. Narejo, M.A. Zaki, et al., Generalized epileptic seizure prediction using machine learning method. Int. J. Adv. Comput. Sci. Appl. **14**(1) (2023)
22. H. Anwar, Q.U. Khan, N. Nadeem, I. Pervaiz, M. Ali, F.F. Cheema, Epileptic seizures. Discoveries **8**(2) (2020)
23. U.R. Acharya, S.V. Sree, G. Swapna, R.J. Martis, J.S. Suri, Automated EEG analysis of epilepsy: a review. Knowl.-Based Syst. **45**, 147–165 (2013)
24. G.C. against Epilepsy, P. for Neurological Diseases, I. B. of Epilepsy, and I. L. against Epilepsy, *Atlas: Epilepsy Care in the World* (World Health Organization, Geneva, 2005)
25. M. Mula, F. Monaco, Ictal and peri-ictal psychopathology. Behav. Neurol. **24**, 21–25 (2011)
26. S.M. Varnosfaderani, R. Rahman, N.J. Sarhan, L. Kuhlmann, E. Asano, A. Luat, M. Alhawari, A two-layer LSTM deep learning model for epileptic seizure prediction, in *2021 IEEE 3rd International Conference on Artificial Intelligence Circuits and Systems (AICAS)* (2021), pp. 1–4
27. R.D. Nass, R. Sassen, C.E. Elger, R. Surges, The role of postictal laboratory blood analyses in the diagnosis and prognosis of seizures. Seizure **47**, 51–65 (2017)
28. R.K. Maganti, P. Rutecki, EEG and epilepsy monitoring. Continuum: Lifelong Learn. Neurol. **19**(3), 598–622 (2013)
29. J.F. Téllez-Zenteno, L.H. Ronquillo, F. Moien-Afshari, S. Wiebe, Surgical outcomes in lesional and non-lesional epilepsy: a systematic review and meta-analysis. Epilepsy Res. **89**(2–3), 310–318 (2010)
30. M. Brázdil, M. Mikl, P. Chlebus, M. Pažourková, Z. Novák, J. Chrastina, J. Prášek, K. Bolčák, D. Fojtíková, J. Horkỳ, et al., Combining advanced neuroimaging techniques in presurgical workup of non-lesional intractable epilepsy. Epileptic Disord. **8**(3), 190–194 (2006)
31. F. Cendes, Neuroimaging in investigation of patients with epilepsy. Continuum: Lifelong Learn. Neurol. **19**(3), 623–642 (2013)
32. R.S. Fisher, J.H. Cross, J.A. French, N. Higurashi, E. Hirsch, F.E. Jansen, L. Lagae, S.L. Moshé, J. Peltola, E. Roulet Perez, et al., Operational classification of seizure types by the international league against epilepsy: position paper of the ILAE commission for classification and terminology. Epilepsia **58**(4), 522–530 (2017)
33. I.E. Scheffer, S. Berkovic, G. Capovilla, M.B. Connolly, J. French, L. Guilhoto, E. Hirsch, S. Jain, G.W. Mathern, S.L. Moshé, et al., ILAE classification of the epilepsies: position paper of the ILAE commission for classification and terminology. Epilepsia **58**(4), 512–521 (2017)
34. P.N. Banerjee, D. Filippi, W.A. Hauser, The descriptive epidemiology of epilepsy—a review. Epilepsy Res. **85**(1), 31–45 (2009)
35. C.M. Michel, D. Brunet, EEG source imaging: a practical review of the analysis steps. Front. Neurol. **10**, 325 (2019)
36. J.A. French, Refractory epilepsy: clinical overview. Epilepsia **48**, 3–7 (2007)
37. S. Wiebe, W.T. Blume, J.P. Girvin, M. Eliasziw, A randomized, controlled trial of surgery for temporal-lobe epilepsy. New England J. Med. **345**(5), 311–318 (2001)

38. I. Ahmad, X. Wang, D. Javeed, P. Kumar, O.W. Samuel, S. Chen, A hybrid deep learning approach for epileptic seizure detection in EEG signals. IEEE J. Biomed. Health Inform. 1–12 (2023)
39. R. Hussein, S. Lee, R. Ward, Multi-channel vision transformer for epileptic seizure prediction. Biomedicines **10**(7), 1551 (2022)
40. I. Azimi, A.M. Rahmani, P. Liljeberg, H. Tenhunen, Internet of things for remote elderly monitoring: a study from user-centered perspective. J. Ambient. Intell. Humaniz. Comput. **8**, 273–289 (2017)
41. L.P. Malasinghe, N. Ramzan, K. Dahal, Remote patient monitoring: a comprehensive study. J. Ambient. Intell. Humaniz. Comput. **10**, 57–76 (2019)
42. B. Yong, Z. Xu, X. Wang, L. Cheng, X. Li, X. Wu, Q. Zhou, IoT-based intelligent fitness system. J. Parallel. Distrib. Comput. **118**, 14–21 (2018)
43. K. Singh, J. Malhotra, Stacked autoencoders based deep learning approach for automatic epileptic seizure detection, in *2018 First International Conference on Secure Cyber Computing and Communication (ICSCCC)* (IEEE, Piscataway, 2018), pp. 249–254
44. K. Singh, S. Singh, J. Malhotra, Spectral features based convolutional neural network for accurate and prompt identification of schizophrenic patients. Proc. Inst. Mech. Eng. H J. Eng. Med. **235**(2), 167–184 (2021)
45. S. L. Oh, V. Jahmunah, N. Arunkumar, E.W. Abdulhay, R. Gururajan, N. Adib, E.J. Ciaccio, K.H. Cheong, U.R. Acharya, A novel automated autism spectrum disorder detection system. Complex Intell. Syst. **7**(5), 2399–2413 (2021)
46. A. Zhang, L. Su, Y. Zhang, Y. Fu, L. Wu, S. Liang, EEG data augmentation for emotion recognition with a multiple generator conditional Wasserstein GAN. Complex Intell. Syst. 1–13 (2021)
47. M. Mc Carthy, P. Schueler, Can digital technology advance the development of treatments for Alzheimer's disease? J. Prev. Alzheimers Dis. **6**(4), 217–220 (2019)
48. K. Singh, J. Malhotra, Cloud based ensemble machine learning approach for smart detection of epileptic seizures using higher order spectral analysis. Phys. Eng. Sci. Med. **44**, 313–324 (2021)
49. Y. LeCun, Y. Bengio, G. Hinton, Deep learning. Nature **521**(7553), 436–444 (2015)
50. O. Faust, Y. Hagiwara, T.J. Hong, O.S. Lih, U.R. Acharya, Deep learning for healthcare applications based on physiological signals: a review. Comput. Methods Programs Biomed. **161**, 1–13 (2018)
51. S. Viglione, G. Walsh, Proceedings: epileptic seizure prediction. Electromyogr. Clin. Neurophysiol. **39**(4), 435–436 (1975)
52. Z. Rogowski, I. Gath, E. Bental, On the prediction of epileptic seizures. Biol. Cybern. **42**(1), 9–15 (1981)
53. Y. Salant, I. Gath, O. Henriksen, Prediction of epileptic seizures from two-channel EEG. Med. Biol. Eng. Comput. **36**, 549–556 (1998)
54. W. van Drongelen, S. Nayak, D.M. Frim, M.H. Kohrman, V.L. Towle, H.C. Lee, A.B. McGee, M.S. Chico, K.E. Hecox, Seizure anticipation in pediatric epilepsy: use of Kolmogorov entropy. Pediatr. Neurol. **29**(3), 207–213 (2003)
55. F. Mormann, T. Kreuz, R.G. Andrzejak, P. David, K. Lehnertz, C.E. Elger, Epileptic seizures are preceded by a decrease in synchronization. Epilepsy Res. **53**(3), 173–185 (2003)
56. L. Kuhlmann, P. Karoly, D.R. Freestone, B.H. Brinkmann, A. Temko, A. Barachant, F. Li, G. Titericz Jr., B.W. Lang, D. Lavery, K. Roman, D. Broadhead, S. Dobson, G. Jones, Q. Tang, I. Ivanenko, O. Panichev, T. Proix, M. NÃ¡hlÃk, D.B. Grunberg, C. Reuben, G. Worrell, B. Litt, D.T.J. Liley, D.B. Grayden, M.J. Cook, Epilepsyecosystem.org: crowd-sourcing reproducible seizure prediction with long-term human intracranial EEG. Brain J. Neurol. **141**, 2619–2630 (2018). 30101347 [pmid]
57. M.I. Maturana, C. Meisel, K. Dell, P.J. Karoly, W. D'Souza, D.B. Grayden, A.N. Burkitt, P. Jiruska, J. Kudlacek, J. Hlinka, et al., Critical slowing down as a biomarker for seizure susceptibility. Nat. Commun. **11**(1), 2172 (2020)

58. H. Daoud, M.A. Bayoumi, Efficient epileptic seizure prediction based on deep learning. IEEE Trans. Biomed. Circuits Syst. **13**(5), 804–813 (2019)
59. N.D. Truong, A.D. Nguyen, L. Kuhlmann, M.R. Bonyadi, J. Yang, O. Kavehei, A generalised seizure prediction with convolutional neural networks for intracranial and scalp electroencephalogram data analysis. CoRR, vol. abs/1707.01976 (2017)
60. Kkostas · Tsiouris, V.C. Pezoulas, M. Zervakis, S. Konitsiotis, D.D. Koutsouris, D.I. Fotiadis, A long short-term memory deep learning network for the prediction of epileptic seizures using EEG signals. Comput. Biol. Med. **99**, 24–37 (2018)
61. D. Thara, B. PremaSudha, F. Xiong, Epileptic seizure detection and prediction using stacked bidirectional long short term memory. Pattern Recogn. Lett. **128**, 529–535 (2019)
62. A. Affes, A. Mdhaffar, C. Triki, M. Jmaiel, B. Freisleben, A convolutional gated recurrent neural network for epileptic seizure prediction, in *How AI Impacts Urban Living and Public Health*, ed. by J. Pagán, M. Mokhtari, H. Aloulou, B. Abdulrazak, M.F. Cabrera (Springer International Publishing, Cham, 2019), pp. 85–96
63. Y. Zhang, Y. Guo, P. Yang, W. Chen, B. Lo, Epilepsy seizure prediction on EEG using common spatial pattern and convolutional neural network. IEEE J. Biomed. Health Inf. **24**(2), 465–474 (2020)
64. H. Khan, L. Marcuse, M. Fields, K. Swann, B. Yener, Focal onset seizure prediction using convolutional networks. IEEE Trans. Biomed. Eng. **65**(9), 2109–2118 (2017)
65. R. Hussein, M.O. Ahmed, R. Ward, Z.J. Wang, L. Kuhlmann, Y. Guo, Human intracranial EEG quantitative analysis and automatic feature learning for epileptic seizure prediction (2019)
66. K. Rasheed, J. Qadir, T.J. O'Brien, L. Kuhlmann, A. Razi, A generative model to synthesize EEG data for epileptic seizure prediction. IEEE Trans. Neural Syst. Rehab. Eng. **29**, 2322–2332 (2021)
67. R. Hussein, S. Lee, R. Ward, M.J. McKeown, Epileptic seizure prediction: a semi-dilated convolutional neural network architecture, in *2020 25th International Conference on Pattern Recognition (ICPR)* (IEEE, Piscataway, 2021), pp. 5436–5443
68. S. Gibson, J.W. Judy, D. Markovic, Technology-aware algorithm design for neural spike detection, feature extraction, and dimensionality reduction. IEEE Trans. Neural Syst. Rehabil. Eng. **18**(5), 469–478 (2010)
69. F. Ali, S. El-Sappagh, S.R. Islam, D. Kwak, A. Ali, M. Imran, K.-S. Kwak, A smart healthcare monitoring system for heart disease prediction based on ensemble deep learning and feature fusion. Inf. Fusion **63**, 208–222 (2020)
70. I. Assali, A.G. Blaiech, A.B. Abdallah, K.B. Khalifa, M. Carrère, M.H. Bedoui, CNN-based classification of epileptic states for seizure prediction using combined temporal and spectral features. Biomed. Sig. Process. Control **82**, 104519 (2023)
71. Y. Ma, Z. Huang, J. Su, H. Shi, D. Wang, S. Jia, W. Li, A multi-channel feature fusion CNN-BI-LSTM epilepsy EEG classification and prediction model based on attention mechanism. IEEE Access (2023)
72. X. Lu, A. Wen, L. Sun, H. Wang, Y. Guo, Y. Ren, An epileptic seizure prediction method based on CBAM-3D CNN-LSTM model. IEEE J. Trans. Eng. Health Med. (2023)
73. M. Shahbazi, H. Aghajan, A generalizable model for seizure prediction based on deep learning using CNN-LSTM architecture, in *Proceedings of the 2018 IEEE Global Conference on Signal and Information Processing (GlobalSIP)* (2018), pp. 469–473
74. T. D.K., P. B.G., F. Xiong, Epileptic seizure detection and prediction using stacked bidirectional long short term memory. Pattern Recogn. Lett. **128**, 529–535 (2019)
75. N.D. Truong, A.D. Nguyen, L. Kuhlmann, M.R. Bonyadi, J. Yang, S. Ippolito, O. Kavehei, Convolutional neural networks for seizure prediction using intracranial and scalp electroencephalogram. Neural Netw. **105**, 104–111 (2018)
76. G. Wang, D. Wang, C. Du, K. Li, J. Zhang, Z. Liu, Y. Tao, M. Wang, Z. Cao, X. Yan, Seizure prediction using directed transfer function and convolution neural network on intracranial EEG. IEEE Trans. Neural Syst. Rehabil. Eng. **28**(12), 2711–2720 (2020)

77. S.M. Varnosfaderani, R. Rahman, N.J. Sarhan, M. Alhawari, A self-aware power management model for epileptic seizure systems based on patient-specific daily seizure pattern, in *2023 International Conference on Microelectronics (ICM)* (IEEE, Piscataway, 2023), pp. 91–95
78. A. Shoeb, Application of machine learning to epileptic seizure onset detection and treatment. PhD thesis (2009)
79. CHB-mit scalp EEG database, Physionet.org, [online]. Available: https://physionet.org/content/chbmit/1.0.0/
80. X. Ma, S. Qiu, Y. Zhang, X. Lian, H. He, Predicting epileptic seizures from intracranial EEG using LSTM-based multi-task learning, in *Chinese Conference on Pattern Recognition and Computer Vision (PRCV)* (Springer, Berlin, 2018), pp. 157–167
81. N.V. Chawla, N. Japkowicz, A. Kotcz, Special issue on learning from imbalanced data sets. ACM SIGKDD Explor. Newsl. **6**(1), 1–6 (2004)
82. D. Snyder, K.W. Leyde, Methods and systems for characterizing and generating a patient-specific seizure advisory system, Aug. 28 2008. US Patent App. 12/035,335
83. S.M. Usman, S. Khalid, M.H. Aslam, Epileptic seizures prediction using deep learning techniques. IEEE Access **8**, 39998–40007 (2020)
84. A.S. Zandi, R. Tafreshi, M. Javidan, G.A. Dumont, Predicting epileptic seizures in scalp EEG based on a variational Bayesian Gaussian mixture model of zero-crossing intervals. IEEE Trans. Biomed. Eng. **60**(5), 1401–1413 (2013)
85. S. Zhao, J. Yang, M. Sawan, Energy-efficient neural network for epileptic seizure prediction. IEEE Trans. Biomed. Eng. **69**(1), 401–411 (2021)
86. C. Reuben, P. Karoly, D.R. Freestone, A. Temko, A. Barachant, F. Li, G. Titericz Jr, B.W. Lang, D. Lavery, K. Roman, et al., Ensembling crowdsourced seizure prediction algorithms using long-term human intracranial EEG. Epilepsia **61**(2), e7–e12 (2020)
87. M.J. Cook, T.J. O'Brien, S.F. Berkovic, M. Murphy, A. Morokoff, G. Fabinyi, W. D'Souza, R. Yerra, J. Archer, L. Litewka, et al., Prediction of seizure likelihood with a long-term, implanted seizure advisory system in patients with drug-resistant epilepsy: a first-in-man study. Lancet Neurol. **12**(6), 563–571 (2013)
88. P.J. Karoly, H. Ung, D.B. Grayden, L. Kuhlmann, K. Leyde, M.J. Cook, D.R. Freestone, The circadian profile of epilepsy improves seizure forecasting. Brain **140**(8), 2169–2182 (2017)
89. I. Kiral-Kornek, S. Roy, E. Nurse, B. Mashford, P. Karoly, T. Carroll, D. Payne, S. Saha, S. Baldassano, T. O'Brien, et al., Epileptic seizure prediction using big data and deep learning: toward a mobile system. EBioMedicine **27**, 103–111 (2018)
90. Z. Zhang, K.K. Parhi, Low-complexity seizure prediction from iEEG/sEEG using spectral power and ratios of spectral power. IEEE Trans. Biomed. Circuits Syst. **10**(3), 693–706 (2015)
91. D. Cho, B. Min, J. Kim, B. Lee, EEG-based prediction of epileptic seizures using phase synchronization elicited from noise-assisted multivariate empirical mode decomposition. IEEE Trans. Neural Syst. Rehab. Eng. **25**(8), 1309–1318 (2016)
92. T. Dissanayake, T. Fernando, S. Denman, S. Sridharan, C. Fookes, Deep learning for patient-independent epileptic seizure prediction using scalp EEG signals. IEEE Sens. J. **21**(7), 9377–9388 (2021)
93. R. Jana, I. Mukherjee, Deep learning based efficient epileptic seizure prediction with EEG channel optimization. Biomed. Sig. Proces. Control **68**, 102767 (2021)
94. Y. Li, Y. Liu, Y.-Z. Guo, X.-F. Liao, B. Hu, T. Yu, Spatio-temporal-spectral hierarchical graph convolutional network with semisupervised active learning for patient-specific seizure prediction. IEEE Trans. Cybernet. **52**(11), 12189–12204 (2021)
95. S.M. Usman, S. Khalid, Z. Bashir, Epileptic seizure prediction using scalp electroencephalogram signals. Biocybern. Biomed. Eng. **41**(1), 211–220 (2021)
96. X. Yang, J. Zhao, Q. Sun, J. Lu, X. Ma, An effective dual self-attention residual network for seizure prediction. IEEE Trans. Neural Syst. Rehabil. Eng. **29**, 1604–1613 (2021)
97. T. Dissanayake, T. Fernando, S. Denman, S. Sridharan, C. Fookes, Geometric deep learning for subject independent epileptic seizure prediction using scalp EEG signals. IEEE J. Biomed. Health Inf. **26**(2), 527–538 (2021)

98. Y. Gao, X. Chen, A. Liu, D. Liang, L. Wu, R. Qian, H. Xie, Y. Zhang, Pediatric seizure prediction in scalp EEG using a multi-scale neural network with dilated convolutions. IEEE J. Trans. Eng. Health Med. **10**, 1–9 (2022)
99. X. Zhang, H. Li, Patient-specific seizure prediction from scalp EEG using vision transformer, in *2022 IEEE 6th Information Technology and Mechatronics Engineering Conference (ITOEC)*, vol. 6 (IEEE, Piscataway, 2022), pp. 1663–1667
100. S.M. Varnosfaderani, I. McNulty, N.J. Sarhan, W. Abood, M. Alhawari, An efficient epilepsy prediction model on European dataset with model evaluation considering seizure types. IEEE J. Biomed. Health Inform. **28**(10), 5842–5854 (2024)
101. Guideline twelve: Guidelines for long-term monitoring for epilepsy. J. Clin. Neurophysiol. **25**(3) (2008)
102. M.R. Sperling, Clinical challenges in invasive monitoring in epilepsy surgery. Epilepsia **38**(s4), S6–S12 (1997)
103. T. Ball, M. Kern, I. Mutschler, A. Aertsen, A. Schulze-Bonhage, Signal quality of simultaneously recorded invasive and non-invasive EEG. NeuroImage **46**(3), 708–716 (2009)
104. E. Behrens, J. Zentner, D. van Roost, A. Hufnagel, C. Elger, J. Schramm, Subdural and depth electrodes in the presurgical evaluation of epilepsy. Acta Neurochir. **128**(1–4), 84–87 (1994)
105. B.H. Brinkmann, J. Wagenaar, D. Abbot, P. Adkins, S.C. Bosshard, M. Chen, Q.M. Tieng, J. He, F. J. Muñoz-Almaraz, P. Botella-Rocamora, J. Pardo, F. Zamora-Martinez, M. Hills, W. Wu, I. Korshunova, W. Cukierski, C. Vite, E.E. Patterson, B. Litt, G.A. Worrell, Crowdsourcing reproducible seizure forecasting in human and canine epilepsy. Brain J. Neurol. **139**, 1713–1722 (2016). 27034258 [pmid]
106. K.M. Tsiouris, V.C. Pezoulas, M. Zervakis, S. Konitsiotis, D.D. Koutsouris, D.I. Fotiadis, A long short-term memory deep learning network for the prediction of epileptic seizures using EEG signals. Comput. Biol. Med. **99**, 24–37 (2018)
107. R.G. Andrzejak, K. Lehnertz, F. Mormann, C. Rieke, P. David, C.E. Elger, Indications of nonlinear deterministic and finite-dimensional structures in time series of brain electrical activity: Dependence on recording region and brain state. Phys. Rev. E **64**, 061907 (2001)
108. A.J. Casson, E. Luna, E. Rodriguez-Villegas, Performance metrics for the accurate characterisation of interictal spike detection algorithms. J. Neurosci. Methods **177**(2), 479–487 (2009)
109. U.R. Acharya, Y. Hagiwara, H. Adeli, Automated seizure prediction. Epilepsy Behav. **88**, 251–261 (2018)
110. Y. Ding, Y. Meng, L. Wang, Study on real-time prediction method of seizures based on yolov3 for EEG spike wave detection, in *Proceedings of the 3rd International Conference on Computer Science and Application Engineering*, CSAE 2019, (New York, NY, USA) (Association for Computing Machinery, New York, 2019)
111. V. Srinivasan, C. Eswaran, N. Sriraam, Approximate entropy-based epileptic EEG detection using artificial neural networks. IEEE Trans. Inf. Technol. Biomed. **11**(3), 288–295 (2007)
112. Emotiv. www.emotiv.com
113. Cadwell. www.cadwell.com/
114. N. Richer, R.J. Downey, W.D. Hairston, D.P. Ferris, A.D. Nordin, Motion and muscle artifact removal validation using an electrical head phantom, robotic motion platform, and dual layer mobile EEG. IEEE Trans. Neural Syst. Rehab. Eng. **28**(8), 1825–1835 (2020)
115. X. Jiang, G.-B. Bian, Z. Tian, Removal of artifacts from EEG signals: a review. Sensors (Basel, Switzerland) **19**, 987 (2019)
116. V. Roy, S. Shukla, Designing efficient blind source separation methods for EEG motion artifact removal based on statistical evaluation. Wirel. Pers. Commun. **108**, 1311–1327 (2019)
117. D. Safieddine, A. Kachenoura, L. Albera, G. Birot, A. Karfoul, A. Pasnicu, A. Biraben, F. Wendling, L. Senhadji, I. Merlet, Removal of muscle artifact from EEG data: comparison between stochastic (ICA and CCA) and deterministic (EMD and wavelet-based) approaches. EURASIP J. Adv. Signal Proces. **2012**, 127 (2012)
118. A. Kilicarslan, J.L.C. Vidal, Characterization and real-time removal of motion artifacts from EEG signals. J. Neural Eng. **16**, 056027 (2019)

119. I. Daubechies, The wavelet transform, time-frequency localization and signal analysis. IEEE Trans. Inf. Theory **36**(5), 961–1005 (1990)
120. A. Hyvärinen, E. Oja, Independent component analysis: algorithms and applications. Neural Netw. **13**(4), 411–430 (2000)
121. P. Flandrin, P. Gonçal'es, G. Rilling, Detrending and denoising with empirical mode decompositions, in *Proceedings of the 2004 12th European Signal Processing Conference* (2004), pp. 1581–1584
122. R.R. Sreekrishna, S. Nalband, A.A. Prince, Real time cascaded moving average filter for detrending of electroencephalogram signals, in *Proceedings of the 2016 International Conference on Communication and Signal Processing (ICCSP)* (2016), pp. 0745–0750
123. P. Gajbhiye, R.K. Tripathy, A. Bhattacharyya, R.B. Pachori, Novel approaches for the removal of motion artifact from EEG recordings. IEEE Sens. J. **19**(22), 10600–10608 (2019)
124. K.T. Sweeney, S.F. McLoone, T.E. Ward, The use of ensemble empirical mode decomposition with canonical correlation analysis as a novel artifact removal technique. IEEE Trans. Biomed. Eng. **60**(1), 97–105 (2013)
125. K.T. Sweeney, H. Ayaz, T.E. Ward, M. Izzetoglu, S.F. McLoone, B. Onaral, A methodology for validating artifact removal techniques for physiological signals. IEEE Trans. Inf. Technol. Biomed. **16**(5), 918–926 (2012)
126. S. Khatun, R. Mahajan, B.I. Morshed, Comparative study of wavelet-based unsupervised ocular artifact removal techniques for single-channel EEG data. IEEE J. Transl. Eng. Health Med. **4**, 1–8 (2016)
127. R. Mahajan, B.I. Morshed, Unsupervised eye blink artifact denoising of EEG data with modified multiscale sample entropy, kurtosis, and Wavelet-ICA. IEEE J. Biomed. Health Inf. **19**(1), 158–165 (2015)
128. S. Yuan, W. Zhou, L. Chen, Epileptic seizure prediction using diffusion distance and Bayesian linear discriminate analysis on intracranial EEG. Int. J. Neural Syst. **28**, 1750043 (2018)
129. G. Geetha, S. Geethalakshmi, EEG de-noising using sure thresholding based on wavelet. Int. J. Comput. Appl. **24**(6), 29–33 (2011)
130. G. Li, S. Jiang, S.E. Paraskevopoulou, M. Wang, Y. Xu, Z. Wu, L. Chen, D. Zhang, G. Schalk, Optimal referencing for stereo-electroencephalographic (sEEG) recordings. NeuroImage **183**, 327–335 (2018)
131. Y. Huang, B. Hajnal, L. Entz, D. Fabó, J.L. Herrero, A.D. Mehta, C.J. Keller, Intracortical dynamics underlying repetitive stimulation predicts changes in network connectivity. J. Neurosci. **39**(31), 6122–6135 (2019)
132. Y. Huang, C. Keller, How can I investigate causal brain networks with iEEG? Preprint. arXiv:2205.07045 (2022)
133. P. Nejedly, J. Cimbalnik, et al., Intracerebral EEG Artifact Identification Using Convolutional Neural Networks. Neuroinform (2018)
134. S. Hu, M. Stead, G.A. Worrell, Automatic identification and removal of scalp reference signal for intracranial EEGs based on independent component analysis. IEEE Trans. Biomed. Eng. **54**(9), 1560–1572 (2007)
135. S.M. Varnosfaderani, I. McNulty, N.J. Sarhan, M. Alhawari, Artifacts removal techniques for the European iEEG dataset, in *2023 IEEE International Midwest Symposium on Circuits and Systems (MWSCAS)* (2023)
136. European Epilepsy Dataset. EPILEPSIAE project (2020). http://epilepsy-database.eu/
137. R. Rahman, S.M. Varnosfaderani, O. Makke, N.J. Sarhan, E. Asano, A. Luat, M. Alhawari, Comprehensive analysis of EEG datasets for epileptic seizure prediction, in *2021 IEEE International Symposium on Circuits and Systems (ISCAS)* (IEEE, Piscataway, 2021), pp. 1–5
138. B. Elahian, M. Yeasin, B. Mudigoudar, J.W. Wheless, A. Babajani-Feremi, Identifying seizure onset zone from electrocorticographic recordings: a machine learning approach based on phase locking value. Seizure **51**, 35–42 (2017)
139. G.A. Worrell, L. Parish, S.D. Cranstoun, R. Jonas, G. Baltuch, B. Litt, High-frequency oscillations and seizure generation in neocortical epilepsy. Brain **127**(7), 1496–1506 (2004)

140. Y. Nagahama, A.J. Schmitt, B.J. Dlouhy, A.S. Vesole, P.E. Gander, C.K. Kovach, D. Nakagawa, M.A. Granner, M.A. Howard, H. Kawasaki, Utility and safety of depth electrodes within the supratemporal plane for intracranial EEG. J. Neurosurg. **131**(3), 772–780 (2018)
141. P. Van Mierlo, B.J. Vorderwülbecke, W. Staljanssens, M. Seeck, S. Vulliémoz, Ictal EEG source localization in focal epilepsy: review and future perspectives. Clin. Neurophysiol. **131**(11), 2600–2616 (2020)
142. P. Van Mierlo, O. Lie, W. Staljanssens, A. Coito, S. Vulliémoz, Influence of time-series normalization, number of nodes, connectivity and graph measure selection on seizure-onset zone localization from intracranial EEG. Brain Topogr. **31**, 753–766 (2018)
143. B.C. Cox, O.A. Danoun, B.N. Lundstrom, T.D. Lagerlund, L.C. Wong-Kisiel, B.H. Brinkmann, EEG source imaging concordance with intracranial EEG and epileptologist review in focal epilepsy. Brain Commun. **3**(4), fcab278 (2021)
144. Y. Yu, X. Si, C. Hu, J. Zhang, A review of recurrent neural networks: LSTM cells and network architectures. Neural Comput. **31**(7), 1235–1270 (2019)
145. S. Hochreiter, Long short-term memory. *Neural Computation* (MIT-Press, Cambridge, 1997)
146. F. Bao, X. Liu, C. Zhang, PyEEG: an open source python module for EEG/MEG feature extraction, in *Computational Intelligence and Neuroscience* (2011)
147. L. Prechelt, Early stopping-but when?, in *Neural Networks: Tricks of the trade* (Springer, Berlin, 2002), pp. 55–69
148. S. West, S.J. Nevitt, J. Cotton, S. Gandhi, J. Weston, A. Sudan, R. Ramirez, R. Newton, Surgery for epilepsy. Cochrane Database Syst. Rev. (6) (2019)
149. R.S. Fisher, J.H. Cross, J.A. French, N. Higurashi, E. Hirsch, F.E. Jansen, L. Lagae, S.L. Moshé, J. Peltola, E. Roulet Perez, et al., Operational classification of seizure types by the international league against epilepsy: Position paper of the ILAE commission for classification and terminology. Epilepsia **58**(4), 522–530 (2017)
150. https://www.epilepsy.com/learn/treating-seizures-and-epilepsy/devices/responsive-neurostimulation-rns
151. L. Iasemidis, Seizure prediction and its applications. Neurodiagn. J. **22**, 489–506 (2011)
152. S. Hooper, E. Biegert, M. Levy, J. Pensock, L. van der Spoel, X. Zhang, T. Zhang, N. Tandon, B. Aazhang, On developing an FPGA based system for real time seizure prediction, in *2017 51st Asilomar Conference on Signals, Systems, and Computers* (2017), pp. 103–107
153. S.M. Kueh, T.J. Kazmierski, Low-power and low-cost dedicated bit-serial hardware neural network for epileptic seizure prediction system. IEEE J. Transl. Eng. Health Med. **6**, 1–9 (2018)
154. C. Tahar, D.I. Maulana, R.R.S. Pandia, T. Adiono, FPGA implementation of deep neural network for wearable pre-seizure detector on epileptic patient, in *2019 16th International Conference on Electrical Engineering/Electronics, Computer, Telecommunications and Information Technology (ECTI-CON)* (2019), pp. 211–213
155. E. Coşgun, A. Çelebi, FPGA based real-time epileptic seizure prediction system. Biocybernet. Biomed. Eng. **41**(1), 278–292 (2021)
156. Z.S. Zaghloul, M. Bayoumi, Early prediction of epilepsy seizures VLSI BCI system. Preprint. arXiv:1906.02894 (2019)
157. C. Lammie, W. Xiang, M.R. Azghadi, Towards memristive deep learning systems for real-time mobile epileptic seizure prediction, in *2021 IEEE International Symposium on Circuits and Systems (ISCAS)* (IEEE, Piscataway, 2021), pp. 1–5
158. P.S. Thorbole, S.D. Kalbhor, V.K. Harpale, V. Bairagi, Hardware implementation of genetic algorithm for epileptic seizure detection and prediction, in *2017 International Conference on Computing, Communication, Control and Automation (ICCUBEA)* (IEEE, Piscataway, 2017), pp. 1–5
159. H. Wang, W. Shi, C.-S. Choy, Hardware design of real time epileptic seizure detection based on STFT and SVM. IEEE Access **6**, 67277–67290 (2018)
160. C. Zhang, M.A.B. Altaf, J. Yoo, Design and implementation of an on-chip patient-specific closed-loop seizure onset and termination detection system. IEEE J. Biomed. Health Inf. **20**(4), 996–1007 (2016)

161. M.A.B. Altaf, C. Zhang, J. Yoo, A 16-channel patient-specific seizure onset and termination detection SoC with impedance-adaptive transcranial electrical stimulator. IEEE J. Solid-State Circuits **50**(11), 2728–2740 (2015)
162. W.-M. Chen, H. Chiueh, T.-J. Chen, C.-L. Ho, C. Jeng, M.-D. Ker, C.-Y. Lin, Y.-C. Huang, C.-W. Chou, T.-Y. Fan, et al., A fully integrated 8-channel closed-loop neural-prosthetic CMOS SoC for real-time epileptic seizure control. IEEE J. Solid-State Circuits **49**(1), 232–247 (2013)
163. C. Tsou, C.-C. Liao, S.-Y. Lee, Epilepsy identification system with neural network hardware implementation, in *2019 IEEE International Conference on Artificial Intelligence Circuits and Systems (AICAS)* (IEEE, Piscataway, 2019), pp. 163–166
164. M. Shoaran, B.A. Haghi, M. Taghavi, M. Farivar, A. Emami-Neyestanak, Energy-efficient classification for resource-constrained biomedical applications. IEEE J. Emerg. Select. Top. Circuits Syst. **8**(4), 693–707 (2018)
165. T. Rukasha, S.I Woolley, T. Kyriacou, T. Collins, Evaluation of wearable electronics for epilepsy: a systematic review. Electronics **9**(6), 968 (2020)
166. P. Lentini, Designing user experience in ehealth applications for young-age epilepsy (2022)
167. S. Tiwari, V. Sharma, M. Mujawar, Y.K. Mishra, A. Kaushik, A. Ghosal, Biosensors for epilepsy management: state-of-art and future aspects. Sensors **19**(7), 1525 (2019)
168. R.U. Hagiwara, S.L. Oh, A deep learning approach for Parkinson's disease diagnosis from EEG signals. Neural Comput. Appl. **22**, 10927–10933 (2020)
169. R. Hussein, M.O. Ahmed, R. Ward, Z.J. Wang, L. Kuhlmann, Y. Guo, Human intracranial EEG quantitative analysis and automatic feature learning for epileptic seizure prediction, CoRR, vol. abs/1904.03603 (2019)
170. M.B. naceur, R. Saouli, M. Akil, R. Kachouri, Fully automatic brain tumor segmentation using end-to-end incremental deep neural networks in MRI images. Comput. Methods Prog. Biomed. **166**, 39–49 (2018)
171. P. Arora, H. Kumar, B.K. Panigrahi, Prediction and analysis of covid-19 positive cases using deep learning models: a descriptive case study of India. Chaos, Solitons Fractals **139**, 110017 (2020)
172. Z. Mei, X. Zhao, H. Chen, W. Chen, Bio-signal complexity analysis in epileptic seizure monitoring: a topic review. Sensors **18**(6) (2018)
173. A. Shawahna, S.M. Sait, A. El-Maleh, FPGA-based accelerators of deep learning networks for learning and classification: a review. IEEE Access **7**, 7823–7859 (2019)
174. J. Liu, H. Zhao, M.A. Ogleari, D. Li, J. Zhao, Processing-in-memory for energy-efficient neural network training: a heterogeneous approach, in *2018 51st Annual IEEE/ACM International Symposium on Microarchitecture (MICRO)* (2018), pp. 655–668
175. Y. Chen, Y. Xie, L. Song, F. Chen, T. Tang, A survey of accelerator architectures for deep neural networks. Engineering **6**(3), 264–274 (2020)
176. H. Al Maharmeh, N.J. Sarhan, C.C. Hung, M. Ismail, M. Alhawari, Compute-in-time for deep neural network accelerators: challenges and prospects, in *2020 IEEE 63rd International Midwest Symposium on Circuits and Systems (MWSCAS)* (2020), pp. 990–993
177. S.E.M.F. Alzghoul et al., A scoring framework and apparatus for epilepsy seizure detection using a wearable belt. J. Med. Signals Sens. **12**(4), 326–333 (2023)
178. N. Chang, T. Chen, C. Chiang, L. Chen, Channel selection for epilepsy seizure prediction method based on machine learning, in *2012 Annual International Conference of the IEEE Engineering in Medicine and Biology Society* (2012), pp. 5162–5165
179. A. Parashar, M. Rhu, A. Mukkara, A. Puglielli, R. Venkatesan, B. Khailany, J. Emer, S.W. Keckler, W.J. Dally, SCNN: an accelerator for compressed-sparse convolutional neural networks. SIGARCH Comput. Archit. News **45**, 27–40 (2017)
180. B. Reagen, P. Whatmough, R. Adolf, S. Rama, H. Lee, S.K. Lee, J.M. Hernández-Lobato, G. Wei, D. Brooks, Minerva: enabling low-power, highly-accurate deep neural network accelerators, in *2016 ACM/IEEE 43rd Annual International Symposium on Computer Architecture (ISCA)* (2016), pp. 267–278

181. F.T. Sun, M.J. Morrell, The RNS system: responsive cortical stimulation for the treatment of refractory partial epilepsy. Expert Rev. Med. Devices **11**(6), 563–572 (2014)
182. N. Sisterson, T.A. Wozny, V.K. et al., A rational approach to understanding and evaluating responsive neurostimulation. Neuroinform **18**, 365–375 (2020)
183. https://www.verywellhealth.com/responsive-neurostimulation-for-epilepsy-4176500
184. M. Shoaran, M. Farivar, A. Emami, Hardware-friendly seizure detection with a boosted ensemble of shallow decision trees, in *2016 38th Annual International Conference of the IEEE Engineering in Medicine and Biology Society (EMBC)* (IEEE, Piscataway, 2016), pp. 1826–1829
185. A. Van de Vel, K. Cuppens, B. Bonroy, M. Milosevic, K. Jansen, S. Van Huffel, B. Vanrumste, P. Cras, L. Lagae, B. Ceulemans, Non-EEG seizure detection systems and potential SUDEP prevention: state of the art: review and update. Seizure **41**, 141–153 (2016)
186. M. Nasseri, E. Nurse, M. Glasstetter, S.Böttcher, N.M. Gregg, A. Laks Nandakumar, B. Joseph, T. Pal Attia, P.F. Viana, E. Bruno, A. Biondi, M. Cook, G.A. Worrell, A. Schulze-Bonhage, M. Dümpelmann, D.R. Freestone, M.P. Richardson, B.H. Brinkmann, Signal quality and patient experience with wearable devices for epilepsy management. Epilepsia, vol. n/a, no. n/a
187. R.G. Andrzejak, K. Lehnertz, F. Mormann, C. Rieke, P. David, C.E. Elger, Indications of nonlinear deterministic and finite-dimensional structures in time series of brain electrical activity: dependence on recording region and brain state. Phys. Rev. E **64**, 061907 (2001)
188. B.H. Brinkmann, J. Wagenaar, D. Abbot, P. Adkins, S.C. Bosshard, M. Chen, Q.M. Tieng, J. He, F.J. Muñoz-Almaraz, P. Botella-Rocamora, J. Pardo, F. Zamora-Martinez, M. Hills, W. Wu, I. Korshunova, W. Cukierski, C. Vite, E.E. Patterson, B. Litt, G.A. Worrell, Crowdsourcing reproducible seizure forecasting in human and canine epilepsy. Brain **139**, 1713–1722 (2016)
189. S.-A. Huang, K.-C. Chang, H.-H. Liou, C.-H. Yang, A 1.9-mw SVM processor with on-chip active learning for epileptic seizure control. IEEE J. Solid-State Circuits **55**(2), 452–464 (2019)
190. F. Forooghifar, A. Aminifar, D. Atienza, Resource-aware distributed epilepsy monitoring using self-awareness from edge to cloud. IEEE Trans. Biomed. Circuits Syst. **13**(6), 1338–1350 (2019)
191. F. Forooghifar, A. Aminifar, D.A. Alonso, Self-aware wearable systems in epileptic seizure detection, in *2018 21st Euromicro Conference on Digital System Design (DSD)* (IEEE, Piscataway, 2018), pp. 426–432

Glossary

Accuracy A metric that indicates how often the model makes correct predictions. It measures the proportion of total correct predictions (both preictal and interictal samples) out of all predictions.

Artifact Unwanted signals or distortions in EEG data caused by external factors such as muscle movement, blinking, or environmental noise. Artifact removal is critical for improving model reliability.

AUC (Area Under the Curve) A measure of classification performance that reflects the model's ability to distinguish between preictal and interictal samples. It represents the area under the ROC curve.

CNN (Convolutional Neural Network) A deep learning model often used for image and signal processing, including feature extraction from EEG data for seizure prediction.

Denoising The process of removing noise and artifacts from EEG signals to enhance signal quality and improve prediction accuracy.

Deep Learning A subset of machine learning that uses neural networks with multiple layers to learn complex patterns from data.

EEG (Electroencephalogram) A test that records electrical activity in the brain using electrodes on the scalp. EEG is a primary tool for diagnosing and predicting epileptic seizures.

Epilepsy A chronic neurological disorder characterized by recurrent, unprovoked seizures due to abnormal brain activity.

Epilepsy Prediction The use of computational models and EEG data to forecast the likelihood of an upcoming seizure, enabling proactive care.

FPR (False Prediction Rate) The frequency at which a model incorrectly predicts a seizure when there is none. A low FPR is important for minimizing false alarms.

Hardware Design The process of creating physical components (such as ASICs or FPGAs) to run machine learning models efficiently for real-time seizure detection and prediction.

S. M. Varnosfaderani et al., *Hardware Software Co-design of Epileptic Seizure Prediction Systems*, Analog Circuits and Signal Processing,
https://doi.org/10.1007/978-3-031-95357-6

iEEG (Intracranial EEG) EEG signals recorded using electrodes placed inside the skull, offering high-resolution data for precise localization of seizure onset zones.

Ictal The phase during a seizure when abnormal electrical activity is occurring in the brain. EEG patterns typically show clear seizure activity in this period.

Interictal The period between seizures when no seizure activity is present. Interictal EEG may still show abnormalities such as spikes or sharp waves.

LSTM (Long Short-Term Memory) A specialized type of recurrent neural network that is well suited for modeling time-series data like EEG due to its ability to retain long-term dependencies.

Machine Learning (ML) A branch of artificial intelligence that allows models to learn patterns from data and make predictions. In epilepsy research, ML is applied to detect and predict seizures.

Postictal The recovery phase after a seizure ends. It is often characterized by confusion, fatigue, or other neurological symptoms and distinctive EEG patterns.

Preictal The phase before a seizure begins. This period is key for seizure prediction, as EEG changes may signal the onset of an impending seizure.

ROC Curve (Receiver Operating Characteristic Curve) A graphical representation that illustrates the performance of a classification model by plotting the true positive rate against the false positive rate at various thresholds.

Seizure Onset Zone (SOZ) The region in the brain where seizures originate. Accurate identification of the SOZ is critical for diagnosis and treatment, including surgical interventions.

Seizure Prediction The task of using algorithms to forecast the occurrence of a seizure before it happens, typically using EEG data.

Sensitivity (Recall) A measure of a model's ability to correctly identify actual seizure events. High sensitivity reduces the likelihood of missed seizures.

TP (True Positive) A correct seizure prediction—when the model predicts a seizure and one actually occurs.

FN (False Negative) A missed seizure—when the model fails to predict a seizure that actually happens.

Index

S. M. Varnosfaderani et al., *Hardware Software Co-design of Epileptic Seizure Prediction Systems*, Analog Circuits and Signal Processing,
https://doi.org/10.1007/978-3-031-95357-6

9783031953569